INTRODUCING ELECTRONIC SYSTEMS

M.W.Brimicombe M.A, D.Phil

Nelson

Thomas Nelson and Sons Ltd
Nelson House   Mayfield Road
Walton-on-Thames Surrey
KT12 5PL   UK

51 York Place
Edinburgh
EH1 3JD   UK

Thomas Nelson (Hong Kong) Ltd
Toppan Building 10/F
22A Westlands Road
Quarry Bay   Hong Kong

Thomas Nelson Australia
480 La Trobe Street
Melbourne   Victoria 3000
Australia

Nelson Canada
1120 Birchmount Road
Scarborough   Ontario
M1K 5G4   Canada

© M.W. Brimicombe 1987

First published by Thomas Nelson and Sons Ltd 1987

ISBN 0-17-448068-7

NPN 9 8 7 6 5

Printed and bound in Hong Kong

Contents

Section E Counters, clocks, controllers and converters

Section F Wireless communications

Introduction

Introducing Electronic Systems is a textbook designed to meet the requirements of GCSE Electronics syllabuses. It assumes that you have no prior knowledge of the subject. Each of the thirty short chapters has a number of questions at the end. These will test your understanding of the material in that chapter. Answers are provided at the back of the book.

Each of the six major sections ends with a number of revision questions. These are the type of questions that you might meet in an electronics exam. They will test your understanding of the material in that section.

The emphasis throughout is on circuits and systems which work, are useful and can be made in a school environment. Electronics is meaningless unless you construct and explore circuits for yourself. Although this book contains no guidance for practical work (that is left to a companion volume of copyright-free practical worksheets), you should make sure that you gain plenty of experience of electronics in action.

As well as showing you how electronic systems work, this book attempts to explain and describe its social and economic implications. Although this is done throughout the book, the consequences of electronic technology are summarised in the last chapter.

Electronics is the emerging technology of today and the dominant technology of tomorrow. ***Introducing Electronic Systems*** will help you to cope with that future and make a contribution towards it.

Circuits which can make decisions

The integrated circuits on this printed circuit board are part of a large modern telephone exchange. It controls the flow of information in the form of electrical pulses between telephones connected to the exchange. *(British Telecom)*

1
Using electricity

Electronic systems

An electronic system is something which uses electricity to obtain, alter, store and transmit **information**.

For example, consider a cassette tape recorder. The whole system consists of four major parts, or **blocks**. These are the **microphone**, the **amplifiers**, the **magnetic tape assembly** and the **loudspeaker**. The **block diagram** of figure 1.1 shows how information is fed from one block to the other.

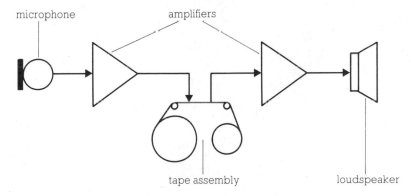

Figure 1.1 The block diagram of a tape recorder

Each block has a different function. The microphone picks up sound waves and converts them into an electrical signal. An amplifier then alters that signal and stores it on the magnetic tape. The other amplifier can, at some later time, read the signal off the tape and feed it into the loudspeaker.

So a cassette tape recorder obtains information about sounds, stores them, and can transmit them at some later time. Obviously, you have no idea as yet about how each of the blocks in the system work. This is not as serious as you might think it is. You can do a lot with electronics if you know what a particular block does, even though you don't know how it does it. This is called the **systems approach** to electronics.

For example, this book is not going to tell you how microphones and loudspeakers work. It will tell you what they do, and how they must be connected to other components to work properly. This is all you need to design a useful electronic device.

Electricity

Electronics is about using electricity to do things. So if you are going to make progress in electronics you need to have some idea of what electricity is and how it behaves. However, you do not need to know the whole complicated truth about electricity in order to do electronics. You will learn enough about the behaviour of electricity for you to be able to understand and design electronic systems. If you want to know more, you will have to seek elsewhere.

A Morse code buzzer

Figure 1.2 shows a simple system which is going to be used to explain what electricity does. It contains three devices and some connecting wires. The wire connects the **power supply**, **buzzer** and **switch** in such a way that the buzzer makes a noise when the switch is pressed.

The system could be used to transmit messages from one place to another using **Morse code**. This is shown below. The switch could be in a different room from the other two components, connected by a pair of wires. By using the Morse code to convert each letter of the alphabet into a series of long and short pressings of the switch, messages could be sent from one room to the other.

A . -		S . . .	
B - . . .		T -	
C - . - .		U . . -	
D - . .		V . . . -	
E .		W . - -	
F . . - .		X - . . -	
G - - .		Y - . - -	
H		Z - - . .	
I . .		0 - - - - -	
J . - - -		1 . - - - -	
K - . -		2 . . - - -	
L . - . .		3 . . . - -	
M - -		4 . . . . -	
N - .		5	
O - - -		6 -	
P . - - .		7 - - . . .	
Q - - . -		8 - - - . .	
R . - .		9 - - - - .	

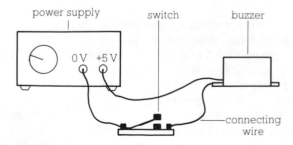

Figure 1.2 A wiring diagram of a Morse code system

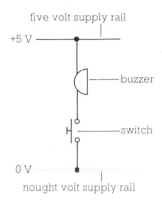

Figure 1.3 A circuit diagram of a Morse code system

Wiring diagrams

Figure 1.2 is an example of a **wiring diagram**. It shows what the system actually looks like when it is assembled. The **circuit diagram** shown in figure 1.3 makes it easier to understand what the electricity is doing in the circuit. Although the circuit diagram looks nothing at all like the real system, it contains all the information that you need to build it.

Circuit diagrams

Each part of the Morse code buzzer system is represented by a **circuit symbol**.

The two horizontal lines of figure 1.3 represent the power supply. They are the **supply rails**. The top one is at +5 V (plus five volts) and the bottom one is at 0 V (nought volts).

The vertical lines represent the **wires** which connect the buzzer and the switch to each other and to the power supply.

Finally, the buzzer and the switch have their own unique symbols. These form part of the international language of electronics. There is a list of other component symbols in the appendix; you ought to have a look at it.

Charge

You can't see electricity, so if you want to describe its behaviour you have to use your imagination. You need a **model** of what electricity is. The model described below is not the only one possible, but it is easy to understand and use.

You have to imagine that metals are a bit like sponges. They have tiny gaps in them, so that the metal is effectively hollow. (The gaps are, of course, far too small to be seen!) Furthermore, you have to imagine that the gaps are filled with something that we will call **charge**. Charge sits in a metal just as water can sit inside a sponge. The charge can move freely around inside the metal, although it cannot escape from it.

The voltage of a bit of metal is a measure of how much energy its charge has.

The charge in the +5 V supply rail has more energy than the charge in the 0 V supply rail. That energy difference will make the charge try to flow from the top supply rail to the bottom one.

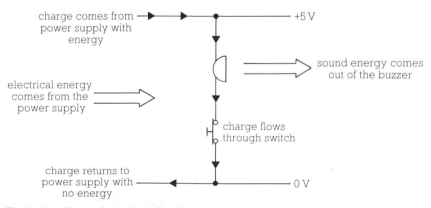

Figure 1.4 Charge flow when the switch is closed

Closed switches

Figure 1.4 shows what happens when the switch is closed. Charge flows from the top supply rail to the bottom one via the buzzer. **The buzzer converts the charge's electrical energy into sound energy.** The charge which enters the bottom supply rail goes into the power supply where it is given back the energy that it lost in going through the buzzer. So the power supply continually feeds energised charge into the top supply rail as fast as it absorbs de-energised charge from the bottom supply rail.

Open switches

When the switch is **open** the buzzer remains silent. As you can see in figure 1.5, the air gap in the switch stops the charge flowing between the supply rails. Air is an **insulator**; charge cannot flow through it. If no charge is moving through the buzzer, then no electrical energy will be converted to sound energy.

A flow of charge in a circuit is called a current.

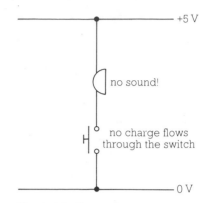

Figure 1.5 Charge flow when the switch is open

Ammeters

Charge is measured in units called **coulombs**. It could take about a minute for one coulomb of charge to flow through a typical buzzer. An instrument called an **ammeter** can be used to detect the flow of charge through a component. The meter reading tells you how much charge is going through in each second. An ammeter measures something which is called the current of a component in units called **amperes** or **amps**.

Thus a current of one amp means that one coulomb of charge is passing through the component in each second. The amp is not a very convenient unit of measurement for electronics. We usually use the **milliamp** instead. **1000 milliamps equals 1 amp.**

Figure 1.6 shows how an ammeter must be connected to measure the current going into a buzzer. Note how the current has to flow through the ammeter before it gets to the buzzer. A typical buzzer might allow a current of 15 milliamps or 15 mA through it.

The current going into a component is always equal to the current coming out of it.

So if you insert the ammeter after the buzzer, as shown in figure 1.7, it will still read 15 mA.

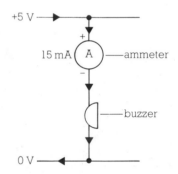

Figure 1.6 Using an ammeter to measure how much current goes into a component

A deaf person's Morse system

The buzzer which we have been discussing is an example of an **output transducer**. These convert electrical energy into other forms of energy. The buzzer produces sound energy each time that the switch is closed. Such a system is clearly useless for communicating with deaf people. For them, a different transducer is required such as a **light bulb**.

Figure 1.8 shows a suitable Morse code system for the deaf. Whenever current goes through it, the bulb emits light energy.

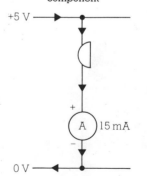

Figure 1.7 Using an ammeter to measure how much current comes out of a component.

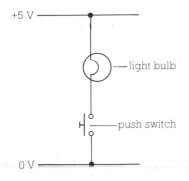

Figure 1.8 A Morse code system for the deaf

Ratings

When choosing a bulb to use in this circuit, you must be careful to select one with the correct **voltage rating**. Since it is destined to operate from a 5 V supply, the bulb must be rated at 5 V. Otherwise it will be either too dim or burn itself out! In fact, a bulb has both a current rating and a voltage rating; the one used in figure 1.8 is rated at 5 V, 60 mA. So when it is connected to a 5 V power supply a current of 60 mA goes through it.

Parallel connection

Suppose that you wanted to design a Morse code transmission system which used both sound and light. The switch would have to control the current through both the bulb and the buzzer. If each component is rated at 5 V they will have to be connected in **parallel**, as shown in figure 1.9.

When the switch is open no current can go through it. So no sound will come from the buzzer or light from the bulb.

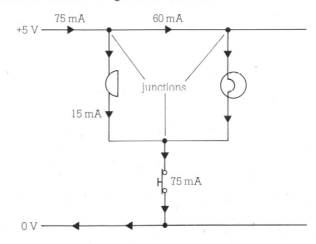

Figure 1.9 Connecting components in parallel

Once the switch is closed current can go through both transducers. So 15 mA will go through the buzzer and 60 mA through the bulb. Both of these currents will have to go through the switch. Therefore a current of 15 + 60 = 75 mA goes through the switch.

> **The sum of the currents going towards a junction equals the sum of the currents leaving it.**

QUESTIONS

1 Look at the circuit in figure 1.10. Which switches have to be pressed so that
   a) only the bulb lights
   b) only the buzzer makes a noise?

2 A motor which is rated at 6 V, 500 mA and a bulb which is rated at 6 V, 50 mA must be connected to the same six volt power supply.
   a) Draw a circuit diagram to show how they must be connected.
   b) How much current will come in and out of the power supply terminals?

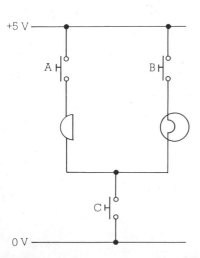

Figure 1.10 Question 1

3 The circuit of figure 1.11 contains a buzzer, a bulb and some ammeters. If the readings of two of those meters are as shown, what do the others read?

4 Copy and complete the following statements.
  a) The two horizontal lines of figure 1.10 are called .... They represent the ....
  b) A buzzer and a bulb are examples of .... They convert .... energy into other forms of energy.
  c) An electric current is a flow of .... through a .... Current is measured in units called .... It can be measured with an .... Current will not flow through ...., and goes from places at high ...... to places at lower .... It will only flow through a switch when it is ....
  d) A bulb is rated at 3 V, 100 mA. To work properly it must be connected to a .... V power supply.

5 A circuit has two switches and a buzzer. The buzzer only makes a noise when both switches are pressed. Draw its circuit diagram.

6 Design a two-way Morse code buzzer system. It has to allow two people to communicate with each other, with messages going both ways between them. Each person will need a switch and a buzzer, but they should be able to share a power supply. Draw a circuit diagram and a wiring diagram of your system.

7 Morse code transmission systems using long lengths of wire to connect the transmitter to the receiver were the first means of rapid long distance communication. They had a large impact on the lives of people when they were first introduced. Explain, with reasons, some of the things which these communication systems allowed people to do that they could not do before.

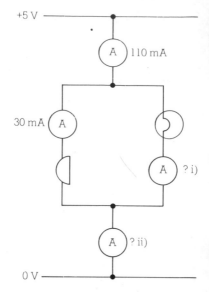

Figure 1.11 Question 3

2
Controlling current

A switch is a very crude way of controlling how much current goes through a transducer. No current when the switch is open. Lots of current when the switch is closed. This chapter is going to explain how **resistors** can be used to give a finer control of current.

Resistors

Resistors are electronic components which obey **Ohm's Law**. This means that the amount of current which goes through them is fixed by the voltage difference between their ends according to this formula:

$$R = \frac{V}{I}$$

V is the voltage across the resistor.
I is the current going through it.
R is the resistance of the resistor.

Resistance is measured in units called **ohms**, usually shortened to Ω. **The resistance of a resistor is supposed to remain constant regardless of how much current flows through it.**

Here is an example of the use of Ohm's Law. Consider the circuit of figure 2.1. It shows a 100 Ω resistor connected to a 5 V power supply. We want to calculate how much current goes through it.

Figure 2.1 A resistor

$$R = \frac{V}{I}$$

$R = 100\ \Omega$
$V = 5\ V$
$I = ?$

therefore $100 = \dfrac{5}{I}$

therefore $I = \dfrac{5}{100}$

therefore $I = 0.05$ A

So a current of 0.05 A goes through the resistor. It is usually more convenient to measure currents in milliamps (**mA**) for electronics. Thus 0.05 A is $0.05 \times 1000 = 50$ mA. Similarly, it is more convenient to quote resistance values in **kilohms** (**kΩ**) rather than in ohms; **1 kilohm is 1000 ohms**.

> **If Ohm's Law is used with R in kilohms and V in volts then I will be in milliamps.**

For example, how much current would flow through a 22 kΩ resistor if it was connected to a 5 V power supply?

$$R = \frac{V}{I}$$

$R = 22$ kΩ
$V = 5$ V
$I = ?$

therefore $22 = \frac{5}{I}$

therefore $I = \frac{5}{22}$

therefore $I = 0.23$ mA

Notice how by making the resistance larger we have made the current smaller. The size of resistor chosen fixes how much current goes between the supply rails.

Colour coding

Resistors are available with a huge variety of resistance values ranging from a few tenths of an ohm to tens of millions of ohms. Since they are very small, it is standard practice to use a series of coloured rings to denote the value of a resistor rather than a series of numbers. The type of resistor you are most likely to use has four coloured rings on it, as shown in figure 2.2.

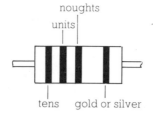

Figure 2.2 The coloured bars on a resistor

Colour	Number
Black	0
Brown	1
Red	2
Orange	3
Yellow	4
Green	5
Blue	6
Purple	7
Grey	8
White	9

Each colour represents a different number. One of the end bands will be gold or silver. To read off the value of the resistor, that band must be at its right hand end. The first two bands on the left then represent a number between 10 and 99, according to the **colour code** shown above. The third band represents the number of noughts which have to be put on the end of the number to get the resistance value in ohms.

For example, suppose that the bands were yellow, purple, red and gold. The yellow and purple bands give the number 47 according to the table above. The red band tells us to add two noughts to this. So the value of the resistor is 4700 Ω or 4.7 kΩ.

What would a 68 kΩ resistor look like ? Its resistance is 68000 Ω. A blue band followed by a grey band represents the number 68. The three noughts after it are represented by an orange band. So a 68 kΩ resistor would have the following bands on it; blue, grey, orange and gold.

Tolerance

The gold or silver band on a resistor tells you what its **tolerance** is. The actual value of a resistor will never be exactly the same as the value coded on it in coloured bands. This is because they are mass produced cheaply. If the tolerance band is gold, then the actual value of the resistor will not be more than 5% larger or smaller than its stated value. A silver band means that the tolerance is 10%.

Preferred values

Only a limited number of resistance values are made. The **preferred values** for the number given by the first two coloured bands are listed in the table below.

10	12	15	18
22	27	33	39
47	56	68	82

Current limiting with resistors

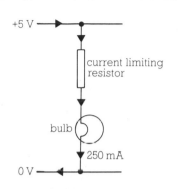

+5 V ➤

current limiting resistor

bulb

▼ 250 mA

0 V ➤

Figure 2.3 Controlling current with a resistor

Resistors can be used to control how much current flows through a transducer. For example, figure 2.3 shows a resistor controlling the current which goes through a light bulb. The bulb is rated at 1.25 V, 250 mA. What value of resistor do we need to place in series with the bulb to operate it off a 5 V supply?

The current going through the resistor will be 250 mA. One end of the resistor will be at +5 V, the other at +1.25 V. So the voltage across the resistor will be 5 − 1.25 = 3.75 V.

We can now use Ohm's law to calculate the resistance.

$$R = \frac{V}{I} \qquad \begin{array}{l} R = ? \\ V = 3.75 \text{ V} \\ I = 250 \text{ mA} \end{array} \qquad \begin{array}{l} \text{therefore } R = \dfrac{3.75}{250} \\ \text{therefore } R = 0.015 \text{ k}\Omega \end{array}$$

So a 15 Ω resistor will do the job.

Power ratings

Resistors convert electrical energy into heat energy. When current goes through a resistor it heats up. This cannot be avoided. A well designed circuit will only generate the minimum amount of heat in its resistors, but some heat will always be generated. The trick is to use resistors which can get rid of the heat without their temperature having to rise too much. Bulky resistors run cooler than small slim ones.

The amount of heat generated in a resistor per second is calculated with this formula.

$$\boldsymbol{W = VI} \qquad \begin{array}{l} W \text{ is the heat generated per second.} \\ V \text{ is the voltage drop across the resistor.} \\ I \text{ is the current which goes through it.} \end{array}$$

The most convenient units for W (the **power**) for electronics is the milliwatt or mW. If **V is measured in volts and I is measured in milliamps then W will be measured in milliwatts.**

As an example, consider the resistor in figure 2.3. How much heat will be generated in it?

$$W = VI \qquad \begin{array}{l} W = ? \\ V = 3.75 \text{ V} \\ I = 250 \text{ mA} \end{array} \qquad \begin{array}{l} \text{therefore } W = 3.75 \times 250 \\ \text{therefore } W = 938 \text{ mW} \end{array}$$

Resistors are available with the following **power ratings.**

250 mW	500 mW	1000 mW	2000 mW

If we used a 250 mW resistor in figure 2.3 it would overheat, get smelly and probably burn out. On the other hand, a 2000 mW resistor would be unecessarily bulky and expensive. A 1000 mW resistor would be suitable; it would get warm, but not so hot that it would get damaged.

Light emitting diodes

Light emitting diodes (LEDs) are transducers which emit light when current goes through them. Unlike light bulbs (which also emit light) they are **solid-state** devices. This means that they have no moving parts and are made out of a specially treated crystal embedded in a coating of translucent plastic. They have a much longer lifetime than bulbs and require far less current. LEDs are coloured; they can be red, yellow or green. Since they work in a completely different way from light bulbs LEDs do not get hot.

Figure 2.4 shows an LED connected to a 5 V power supply. The current flows from the **anode** of the LED to its **cathode** and it is controlled by the resistor. All LEDs are rated at about 2 V, so a resistor must always be connected in **series** with them.

How large should the resistor be? Suppose that the LED is rated at 2 V, 6 mA. We can then use Ohm's law to calculate a suitable value for the resistor. (Figure 2.5.)

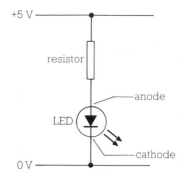

Figure 2.4 Connecting an LED to a power supply

$$R = \frac{V}{I} \qquad \begin{aligned} R &= ? \\ V &= 5 - 2 = 3 \text{ V} \\ I &= 6 \text{ mA} \end{aligned} \qquad \begin{aligned} \text{therefore } R &= \frac{3}{6} \\ \text{therefore } R &= 0.5 \text{ k}\Omega \end{aligned}$$

So a 0.5 kΩ or 500 Ω resistor would do. The nearest preferred value is 470 Ω. As well as considering its resistance, we need to think about its power rating as well. So we have to calculate the power of the resistor.

$$W = VI \qquad \begin{aligned} W &= ? \\ V &= 3 \text{ V} \\ I &= 6 \text{ mA} \end{aligned} \qquad \begin{aligned} \text{therefore } W &= 3 \times 6 \\ \text{therefore } W &= 18 \text{ mW} \end{aligned}$$

We can obviously choose a 250 mW resistor without worrying about how hot it will get!

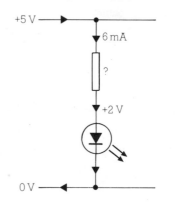

Figure 2.5 Choosing a suitable current limiting resistor

LED characteristics

The graph of figure 2.6 is the ***I–V characteristic*** of a typical LED. Notice how the current climbs rapidly once the voltage exceeds 2 V.

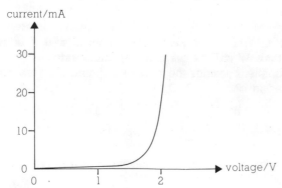

Figure 2.6 The *I–V* characteristic for a typical LED

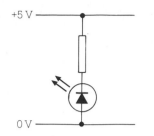

Figure 2.7 A reverse biased LED

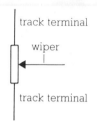

Figure 2.8 A potentiometer

LEDs will only work if they are **forward biased**. The current must go from anode to cathode. If you try to make the current flow from cathode to anode (as shown in figure 2.7) the LED will be **reverse biased** and will not work. Indeed, no current will flow though it.

Like all electronic components, LEDs can be destroyed if the current or voltage gets too great. A typical LED cannot survive if the current exceeds 50 mA or the reverse bias voltage exceeds 5 V.

Potentiometers

A resistor has a fixed resistance. A variable resistor can be made from a component called a **potentiometer**. Its circuit symbol is shown in figure 2.8. If you use the two **track terminals** the potentiometer behaves just like a resistor. However, if you use one of the track terminals and the **wiper**, the resistance of the device can be changed by rotating the knob of the potentiometer.

For example, look at figure 2.9. The potentiometer is being used as a variable resistor, so that the current going through the LED can be altered. The resistance between the track terminals is 5 kΩ. The resistance between the wiper and a track terminal can be varied between 0 kΩ and 5 kΩ by rotating the potentiometer knob.

The 100 Ω resistor is in the circuit so that there is always some resistance present to limit the current going through the LED. So the potentiometer allows the total resistance in the circuit to be varied between 100 Ω and 5100 Ω.

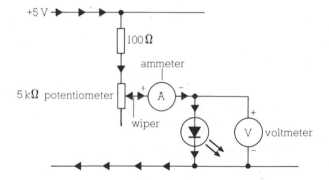

Figure 2.9 Using a potentiometer to vary the current going through an LED

Ammeters and voltmeters

Take another look at figure 2.9. The ammeter measures how much current goes into the LED. There is also a **voltmeter** which measures the voltage drop across the LED. Although voltmeters tend to look very much like ammeters, they are used in very different ways.

An ammeter is always connected in series with a component. **A good ammeter will have a negligible resistance**. This is so that it does not alter the current which goes through a circuit when it is inserted.

On the other hand, a voltmeter is always connected in parallel with a component, and will measure the voltage difference between the ends of the component. **A good voltmeter will have a very high resistance**. So very little current will go through it when it is connected to a circuit.

QUESTIONS

1 Use Ohm's law to calculate the current which goes though each of the resistors in figure 2.10.

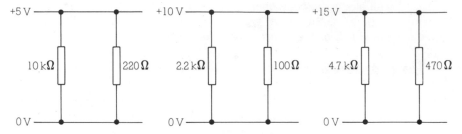

Figure 2.10 Question 1

2 Calculate the power of each of the resistors in figure 2.10. Which of them could safely have a power rating of 250 mW?

3 The motor of figure 2.11 is rated at 3 V, 200 mA. Do calculations to show that a 10 Ω, 500 mW resistor could be used in the circuit.

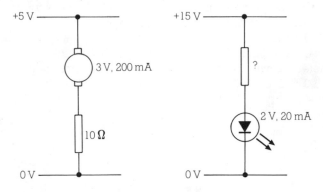

Figure 2.11 Question 3   **Figure 2.12** Question 4

4 Figure 2.12 shows an LED being run off a 15 V supply. If it is rated at 2 V, 20 mA calculate a suitable value for the resistor.

5 A number of circuits are shown in figure 2.13. The ammeter is supposed to measure the current going through the bulb and the voltmeter the voltage drop across the resistor. Which circuit has the meters correctly connected?

6 Give the values of these resistors in ohms and kilohms.
   a) Red, red, orange, gold.
   b) Yellow, purple, brown, silver.
   c) Brown, black, yellow, gold.
   d) Orange, orange, red, silver.

7 Give the coloured bands you would expect to see on the following resistors.
   a) 68 kΩ, 10%
   b) 120 Ω, 5%
   c) 2.2 kΩ, 10 %
   d) 330 Ω, 5%

8 Copy and complete the following statements.
  a) An LED will only glow if the current flows from its . . . . . to its
      . . . . . . The voltage across a lit LED is approximately . . . . . .
  b) The gold band on a resistor represents its . . . . . .
  c) Ammeters are always connected in . . . . . with a component. They
      measure the . . . . . a component. Ammeters have a . . . . . resis-
      tance.
  d) Voltmeters are always connected in . . . . . with a component. They
      measure the . . . . . a component. Voltmeters have a . . . . . resis-
      tance.
  e) A potentiometer has . . . . . terminals. To use it as a variable
      resistor you have to use the . . . . . and one of the . . . . . .

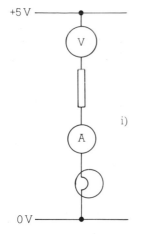

Figure 2.13 Question 5

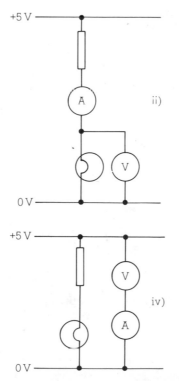

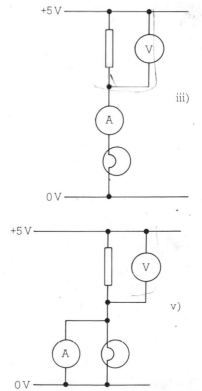

3
Controlling voltage

Resistors are versatile components. As well as controlling current, they can be used to control the voltage at a point in a circuit.

Variable voltage sources

Figure 3.1 shows how a potentiometer has to be connected to make a **variable voltage source**. One track terminal is held at +5 V, the other at 0 V. The voltage of the wiper can be varied from 0 V to +5 V by rotating the knob on the potentiometer. The system allows any voltage between those of the supply rail voltages to be generated.

Potentiometers are widely used as variable voltage sources. They can also be used as **input transducers**, converting the angle of their shaft into a voltage. A typical potentiometer can have its shaft rotated through 300°. So a 1 V change in the voltage of the wiper corresponds to a 60° rotation of the shaft.

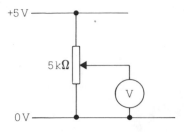

Figure 3.1 A variable voltage source

Fixed voltage sources

By putting two resistors in series you can convert the voltage of a power supply into a smaller voltage.

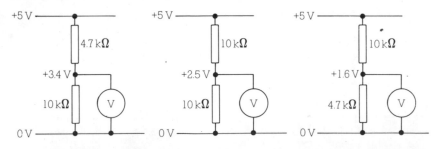

Figure 3.2 Three voltage dividers

For example, look at figure 3.2. It shows three **voltage dividers**. Each pair of resistors produces a voltage at their junction which is then measured by a voltmeter. The value of the voltage produced depends on the ratio of the resistors in the voltage divider.

The two resistors in a voltage divider are given special names. The top one is called the **pull-up resistor** (R_T) and the bottom one is called the **pull-down resistor** (R_B); see figure 3.3. The value of the output voltage V_{OUT} will depend on the relative sizes of R_T and R_B as well as the supply voltage.

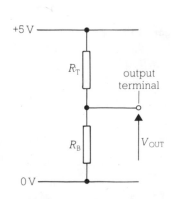

Figure 3.3 A voltage divider

Type of voltage divider	Output voltage V_{OUT}
R_T greater than R_B	about 0 V
R_T equal to R_B	+2.5 V
R_T smaller than R_B	about +5 V

It is useful to refer to any voltage between +2.5 V and +5 V as **high** and any voltage between 0 V and +2.5 V as **low**. Then the behaviour of a voltage divider can be summarised as follows:

V_{OUT} is high if R_B is larger than R_T.
V_{OUT} is low if R_B is smaller than R_T.

Simple processing systems

Voltage dividers are extensively used to interface input transducers to electronic circuits. They can be used to convert information into an electrical signal.

Light dependent resistors

Figure 3.4 shows a voltage divider which has a **light dependent resistor (LDR)** in it. An LDR behaves like a normal resistor, except that its resistance depends on how much light is falling on it. Its resistance is very high in the dark, typically several megohms (**1 megohm = 1000 kilohms**). As the amount of light falling on it increases, the resistance of an LDR drops steadily. It will be less than 1 kΩ in bright light.

A light sensor

The voltage of the output terminal will therefore depend on how much light is hitting the LDR. If there is no light the LDR (the pull-up resistor) will be very much larger than the 10 kΩ resistor. So the output voltage will be low.

But if lots of light hits the LDR then the output voltage will be high. The pull-up resistor will be much smaller than the pull-down resistor.

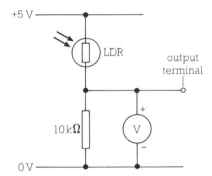

Figure 3.4 A light sensing circuit

Light conditions	Output Voltage
Dark	Low
Bright	High

So the output of the system (a voltage) carries information about how brightly lit the LDR is. A later chapter will show you how that information can be processed electronically.

Water contacts

A circuit which can detect water is shown in figure 3.5. The pull-up resistor of the voltage divider is a pair of contacts. These are simply a pair of bare wires that are close together. Normally there is air between the contacts so their resistance is very large. So V_{OUT} is pulled low by the 100 kΩ pull-down resistor.

If the contacts are placed in water their resistance falls to much less than 100 kΩ. (Water is not a very good conductor, but it does let some

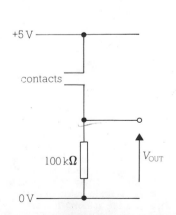

Figure 3.5 A water sensing circuit

current go through it.) This means that the pull-up resistor is much smaller than the pull-down resistor and so V_{OUT} gets pulled up to about $+5$ V.

State of contacts	Output voltage
Dry	Low
Wet	High

Remote sensing

The system could be used to monitor the state of a large water tank. The contacts could be placed in the tank just below the top. The rest of the circuit (the power supply, the pull-down resistor and the voltmeter) could be somewhere else, connected to the contacts by a pair of wires. The system would then tell you if the tank was 'full' or 'not full.' Indeed, you could write 'full' over the 5 V mark and 'not full' over the 0 V mark of the voltmeter!

Clearly, a number of such systems could allow one person to monitor the state of a number of separate water tanks. Those tanks could be far away and in inaccessible locations, so that it might not be easy to monitor them in any other way.

Switches as sensors

You can think of a switch as a special sort of resistor. It has a resistance of 0Ω when it is pressed (closed). When it is not pressed (open) its resistance is enormous. Switches are widely used to interface human fingers to electronic systems.

For example, consider the circuit shown in figure 3.6. V_{OUT} will be low when the switch is open. The 1 kΩ pull-down resistor will be much smaller than the pull-up resistor (the switch). But when the switch is closed V_{OUT} will be pulled high. This is because the pull-up resistor is now very much smaller than the pull-down resistor.

Simple logic gates

Switches and resistors can be used to make simple **logic gates**. As well as converting information into electrical form they can process it.

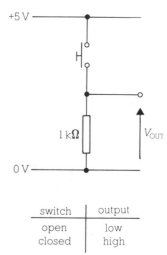

switch	output
open	low
closed	high

Figure 3.6 A touch sensing circuit

Part of the railway signal box at London Bridge station. Sensors placed under the tracks send signals to the display board so that one person can monitor all of the trains near the station. *(British Railways Board)*

switch A	switch B	output
open	open	low
open	closed	low
closed	open	low
closed	closed	high

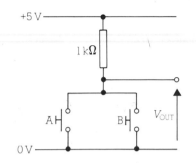

Figure 3.7 A simple logic gate

Figure 3.7 has two switches in series as the pull-up resistor of a voltage divider. The pull-up resistor will only be smaller than 1 kΩ when both switches are pressed. So the output will only go high when both switches are pressed.

Similarly, the output of the circuit in figure 3.8 will only go high if neither of the switches are being pressed. Only then will the pull-down resistor (the two switches in parallel) have a resistance of more than 1 kΩ.

switch A	switch B	output
open	open	high
open	closed	low
closed	open	low
closed	closed	low

Figure 3.8 A simple logic gate

Using a logic gate

What could this type of system be used for? Here is an example. Two astronauts are sitting in a space capsule. They are going over their checklists, ensuring that everything is as it should be. Each astronaut has a switch in front of him. When he is satisfied that everything is correct in his half of the capsule he puts his finger on his switch. The two switches are part of a circuit like the one of figure 3.7, with the power supply, pull-down resistor and voltmeter in the launch control room some distance away. As soon as the voltmeter shows that V_{OUT} is high, launch control will know that both astronauts are ready for take-off.

QUESTIONS

1 Figure 3.9 shows a number of voltage dividers. For which of them will the output voltage be high?

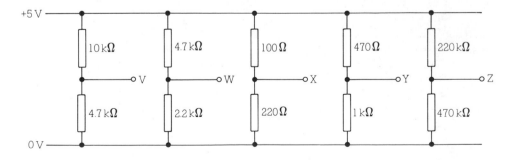

Figure 3.9 Question 1

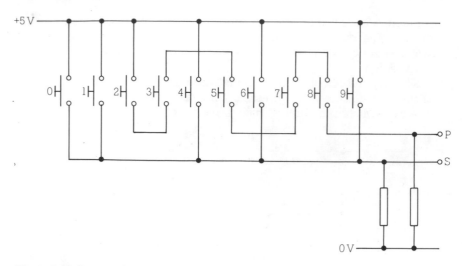

Figure 3.10 Question 2

2 Figure 3.10 shows a simple electronic security system. The ten switches are labelled from 0 to 9. They are the ten inputs to the system. The two outputs are labelled P (for pass) and S (for stop) and are monitored by a pair of voltmeters where they can be seen by the security guards in a building. In order to get into the building you have to press the switches to make P go high without making S go high. Which five switches have to be pressed simultaneously to make P go high?

3 Design a circuit which would tell you that a water tank was empty. Its output must be low if there is water in the tank and high if there is no water in the tank.

4 A circuit contains two LDRs, a pull-down resistor and a voltmeter. The reading on the voltmeter is only high when both of the LDRs are in bright light. Draw a circuit diagram for the system.

5 A circuit contains two switches and a 1 kΩ resistor. Its output terminal must only go low when both switches are pressed. Draw a circuit diagram for the system.

6 Copy and complete the following statements.
   a) When the output terminal of a circuit is between +5 V and +2.5 V it is said to be . . . . . .If it is between +2.5 V and 0 V it is said to be . . . .
   b) Two resistors in series with a power supply make a . . . . . . divider. The resistor connected to the top supply rail is called the . . . . resistor. The resistor connected to the bottom supply rail is called the . . . . resistor.
   c) An LDR has a resistance which depends on . . . . In bright light its resistance is . . . . . In the dark, its resistance is . . . .

4 .

Increasing the current

In the last chapter you met a number of circuits which could make simple decisions. The state of their output terminal (which was high or low) told you something about the state of their inputs. This chapter will show you how **integrated circuits** and **relays** can be used to allow such circuits to control more than just a voltmeter.

A light meter

Figure 4.1 shows a circuit which might be useful in a camera. It is a simple **light level indicator**. The voltage divider on the left contains an LDR. The reading on the voltmeter will depend on the amount of light falling on the LDR. If there is only a little light (not enough to take a good photograph) the LDR will have a big resistance and the voltmeter reading will be high.

Of course, to be really useful, the circuit needs to fit inside the camera, so that the LDR can monitor the amount of light getting through the camera lens. Furthermore, it would be useful if the voltmeter could be seen while you were looking through the camera viewfinder.

Voltmeters are expensive, especially when they are miniaturised. They are not very rugged; they are easily damaged by shock. So another method of monitoring the output of the voltage divider is required. Perhaps we could use an LED as shown in figure 4.2. After all, LEDs are small, cheap and rugged (they have no moving parts which can get jolted out of place), so they would appear to be the perfect solution to the problem.

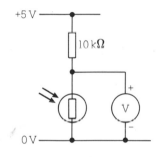

Figure 4.1 A light meter

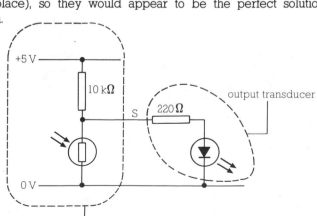

Figure 4.2 A rugged light meter

Loading effects

Unfortunately, the circuit of figure 4.2 doesn't work properly. The **block diagram** of figure 4.3 will help you to understand why. A block diagram shows you how the important component parts of a system are related to each other. The light level indicator of figure 4.2 consists of two main parts. The **signal source** on the left controls the **output transducer** on the right. This is made clear in figure 4.3.

Figure 4.3 A block diagram of the light meter

The signal source is the voltage divider. Its output (marked S) is high in the dark and low in the light. The output transducer is the LED and its current limiting resistor. When its input (S) is high, the LED will glow.

The system does not work because the LED will have a current of about 13 mA going through it when S is at +5 V. That current has to come from the output of the signal source, and you cannot draw 13 mA from the voltage divider without preventing it from working properly. The basic problem with the circuit is that the LED stops the output terminal of the voltage divider from going high when the LDR is in the dark.

Once you start drawing current from the output terminal of a voltage divider it stops working properly.

Limits on current

In general, voltage dividers can only drive voltmeters with their output terminals. Any current which flows out of the output terminal is diverted from the pull-down resistor, causing the voltage across that resistor to be less than it would be otherwise.

So how much current can you draw from a voltage divider without upsetting it? As little as possible is the answer. Certainly much less than the current which is normally going through the two resistors of the voltage divider. Real voltmeters have resistances of more than 100 kΩ, so they draw less than 50 microamps from the output terminal of a voltage divider. (**1000 microamps = 1 milliamp**.)

Buffers

A **buffer** is a device which can boost the current output of a signal source. Its function is obvious if you compare the block diagrams of figure 4.3 and 4.4. The buffer sits between the signal source and the output transducer, and it allows the whole system to work as specified. The addition of the buffer allows the LED to glow when the LDR is in the dark.

The buffer draws very little current from the signal source, so the voltage divider is able to work properly. The point marked A will be high in the dark and low in the light. At the same time, the buffer is able to

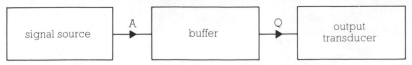

Figure 4.4 A block diagram of a useable light meter

provide lots of current for the output transducer. So the LED will glow brightly when A is high.

The extra current for the output transducer comes from the supply rails which are connected to the buffer. Buffers are **active electronic components** because they will only work if they are connected to a power supply.

The 555 IC

There are many types of buffer available. Figure 4.5 shows how a particular one (known as the **555 IC**) can be used to translate the block diagram of figure 4.4 into a circuit diagram.

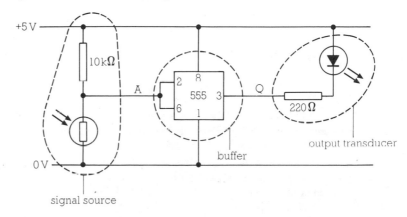

Figure 4.5 The circuit diagram of a useable light meter

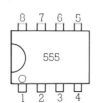

Figure 4.6 Top view of a 555 integrated circuit

The 555 IC is a small black plastic package which has eight metal legs coming out of it, four on each side. It looks a bit like a spider, and is shown in figure 4.6. The legs (or **pins**) allow electrical connections to be made to the circuit which is etched onto a slice of silicon at the centre of the black plastic. It is called an **integrated circuit** (or **IC**) because all of the resistors and transistors which make up the circuit of the buffer are built out of the same crystal of silicon.

Advantages of ICs

The circuit of the buffer is quite complex, with many components. Figure 4.7 shows a circuit which behaves like the 555 IC when it is used as a buffer. The circuit is built out of **discrete components** (resistors and transistors), and is relatively bulky when assembled. The 555 IC is much smaller. Furthermore, circuits which are made from discrete components have to be assembled and soldered together by human beings. **This makes them expensive.**

ICs are very cheap to manufacture if they are made in large numbers, as much of the manufacturing process can be automated, requiring only a little intervention from human beings.

So integrated circuits allow electronics to be cheap, robust and small. It is precisely these qualities which has led to the recent rapid expansion of electronics.

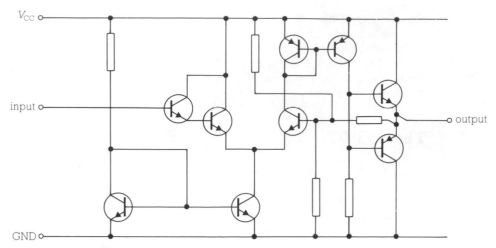

Figure 4.7 A circuit which behaves like a 555 IC buffer

Pin connections

Figure 4.6 shows what the 555 IC looks like when you view it from the top. The pins are numbered from 1 to 8. The block diagram of figure 4.8 shows you what some of the pins are for.

Pins 1 and 8 are used for the power supply connections. Pin 1 must be connected to the bottom supply rail which is usually at 0 V. It is labelled GND (short for **ground**) in figure 4.8. Pin 8 must be connected to the top supply rail, labelled V_{CC}. V_{CC} can have any value from $+5$ V to $+15$ V.

Pin 1 usually has a small dot impressed into the plastic next to it. In any case, it is always at the bottom left hand corner of the IC if it is placed so that you can read the numbers printed on it. **It is vital that you always connect an IC to the power supply correctly**. Wrong connection usually makes the IC self-destruct, producing a lot of smoke and a nasty smell.

Pins 2 and 6 have to be connected together to make the input terminal of the buffer. Pin 3 is then the output. None of the other pins (4, 5 and 7) must be connected to anything.

555 buffer behaviour

The behaviour of a 555 IC as a buffer is shown in figure 4.9. A pair of bulbs have been connected to the output terminal.

When the input is high, the output is low.
When the input is low, the output is high.

So when the input is high (as shown on the top), the output is low. It acts as the bottom supply rail for the top bulb, sitting at about 0 V. Current comes from the $+5$ V supply rail, through the top bulb, into the 555 output and gets delivered into the bottom supply rail.

The bottom of figure 4.9 shows what happens when the input is low. The output goes high, so that current will go through the bottom bulb. That current, of course, comes from the top supply rail via the 555's output, and ends up going into the bottom supply rail.

The current flowing into the input of a 555 is very small.
The output of a 555 can handle currents of up to 100 mA with ease.

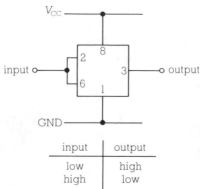

input	output
low	high
high	low

Figure 4.8 A 555 IC buffer

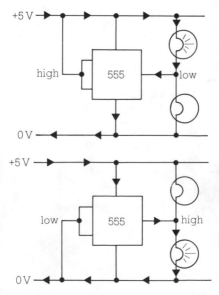

Figure 4.9 Current flow in and out of a buffer's output terminal

Water level alarm

What follows is an example of how a 555 IC can be used to design a useful electronic system.

The first step in any design is to specify the problem. We want the system to send out an alarm signal if the water in a container gets above a certain level.

It then helps to draw a **block diagram**. For this particular problem, the block diagram of figure 4.10 would do. It shows that we need a water sensor to control a sound source via a buffer.

Figure 4.10 Block diagram of a water level alarm system

Now we have to think about suitable circuit elements for each block. The water sensor could be a voltage divider containing a pair of contacts as one of the resistors. The buffer will be a 555 IC, and the sound source could be a buzzer.

Eventually, we end up with the **circuit diagram** of figure 4.11. We have chosen to use the contacts as the pull-down resistor in the voltage divider, so that A goes low in the wet. This means that Q goes high in the wet, sending current through the buzzer, generating the alarm signal.

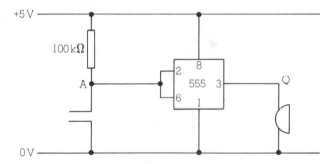

Figure 4.11 Circuit diagram of a water level alarm system

Relays

The 555 IC is a very useful buffer. It can handle output currents of up to 100 mA, allowing it to control LEDs, buzzers and small light bulbs. But it can only control devices which are are rated at, or below 5 V, 100 mA. So it cannot control motors or solenoids as these need too high a voltage or current. To control large currents and voltages, you need to use a **relay**.

The type of relay we are going to consider (and there are many types to choose from) has its circuit symbol shown in figure 4.12. **A relay is basically a switch which is controlled by an electromagnet**. The switch part of the relay has one **pole** and two **throws**, so it is a **s**ingle-**p**ole-**d**ouble-**t**hrow or **SPDT** type.

The state of the switch depends on how much current is going through

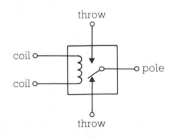

Figure 4.12 An SPDT relay

the relay coil. When no current goes through the coil, the pole is connected to the bottom throw as drawn in figure 4.12. If a current of at least 50 mA goes through the coil then the pole is connected to the top throw instead.

Controlling motors

Figure 4.13 shows how a relay could control a small motor. When both of the coil terminals are at the same voltage, no current goes through the motor. But if the coil terminals are held 5 V apart, enough current goes through the coil to make the relay switch contacts change, so that current can go through the motor. Note that only 70 mA has to go through the coil for 200 mA to go through the motor.

A small current going through the relay coil can control a much larger current going through the relay switch contacts.

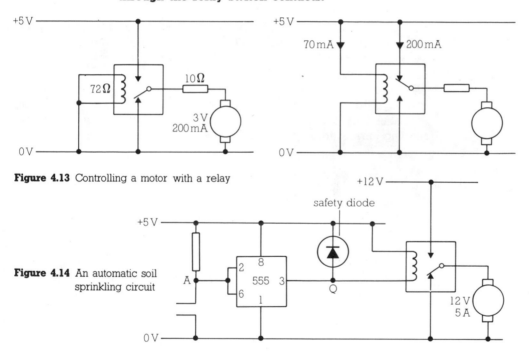

Figure 4.13 Controlling a motor with a relay

Figure 4.14 An automatic soil sprinkling circuit

A soil sprinkler

Figure 4.14 shows how a relay can be used in practice. The whole system is designed to monitor the dampness of the soil in a greenhouse and pump water into that soil if it gets too dry.

The contacts in the voltage divider are buried in the soil. Once the soil is too dry, there will not be enough water between the contacts to keep their resistance low, and A will go high.

The 555 IC acts as a buffer for the voltage divider. When A goes high, Q will go low, pulling a current of about 70 mA through the relay coil. So the relay contacts swap over and the motor rotates. Notice how the motor is rated at 12 V, 5 A, so it needs to run off supply rails at +12 V and 0 V.

The motor drives a pump which squirts water onto the greenhouse soil. When the soil gets damp enough, A will go low again, making Q go high.

As soon as Q goes high, current no longer goes through the relay coil, the relay contacts switch over and disconnect the motor from its power supply. The safety diode is included in the circuit to prevent the 555 IC from being damaged by voltage surges which happen when the current suddenly stops going through the coil. Such diodes must always be connected when ICs drive relay coils.

QUESTIONS

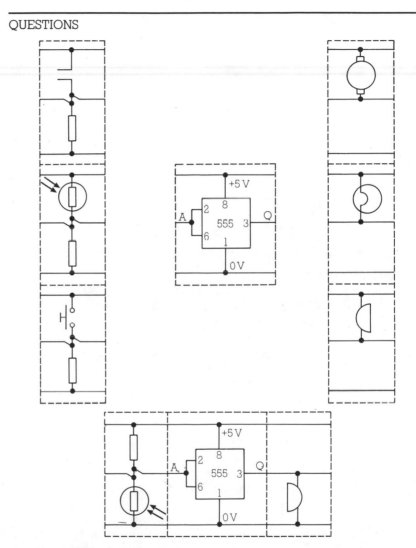

Figure 4.15 Question 1

1   Figure 4.15 shows three types of signal source, one buffer and three types of output transducer.
   a)   Describe the behaviour of the buffer (say how the state of Q depends on the state of A).
   b)   Figure 4.15 shows an example of a complete system. Which one of the following statements describes the behaviour of the system?
      i)   The buzzer only makes a sound in the dark.
      ii)   The buzzer only makes a sound in the light.
      iii) The buzzer makes a sound all of the time.
      iv)   The buzzer never makes a sound.

2 You are going to design some systems using the circuit elements shown in figure 4.15. Bear in mind that the signal sources and output transducers can be used either way up. For each of the systems described below draw

      i) a block diagram,

      ii) a circuit diagram.

a) The motor must only rotate when it gets dark.

b) The bulb only goes off when the switch is pressed.

c) The buzzer makes a noise when the contacts are wetted.

d) The bulb must come on only when it gets dark.

e) The bulb must come on when the contacts are dry.

3 A number of output transducers are listed below, together with their voltage and current ratings. Which ones could be controlled by a 555 IC? Which ones need to be controlled by a relay?

a) A 5 V, 60 mA light bulb.

b) A 12 V, 100 mA light bulb.

c) A 2 V, 40 mA LED

d) A 12 V, 1 A solenoid

e) A 24 V, 25 mA sound source

4 What is the difference between an integrated circuit and one made of discrete components? State, with reasons, two ways in which integrated circuits are superior to ones made of discrete components.

5 You have been asked to design an electronic system which will make a noise whenever light hits an LDR. (The LDR is to be built into the top of a table, underneath a valuable vase. Each time that the vase is removed, the LDR will be exposed to light and the alarm will sound.) You have a siren which is rated at 24 V, 600 mA.

a) Draw a block diagram of a suitable system.

b) Explain how the system will work.

c) Draw a circuit diagram of your system.

The buffer game

This game requires a set of seven cards as shown in figure 4.15. The cards may be put together in groups of three to make circuit diagrams of complete electronic systems. An example is shown in figure 4.15; note how the signal source and output transducer cards can be turned upside down.

The object of the game is to use the cards to make working systems and to challenge the other players to say what those systems do. For example, the system shown in figure 4.15 has the following behaviour; the buzzer only makes a noise when there is lots of light.

Each player in turn uses the cards to make up a circuit. He then challenges one of the other players (any one) to tell him what the circuit does. If that player gives a correct answer he gains a point. If the answer is wrong, then the player can gain a point by correctly explaining what the circuit does.

5
Basic logic gates

The electronic systems which you met in the last chapter contained three types of sub-system.

First of all there was a **signal source**. It consisted of an input transducer (such as an LDR) in a voltage divider. Then there was an **active component** such as a buffer (and perhaps a relay as well) to process the signal from the signal source. Finally there was an **output transducer** (such as a light bulb).

This chapter is going to show how a special class of active components called **logic gates** can be used to combine the inputs from several signal sources to control a single output transducer.

Figure 5.1 A NOT gate

The NOT gate

The simplest of all the logic gates is shown in figure 5.1. It is called a **NOT gate.**. The symbol shown in figure 5.1 has one input (marked A) and one output (marked Q), but there are no power supply connections shown. In practice a NOT gate always needs to be connected to a pair of supply rails, but this is never shown in circuit diagrams. This is because circuits often contain many logic gates connected to each other, and the power supply connections tend to make the circuit diagrams over-complicated.

Logic levels

The graph of figure 5.2 shows the **transfer characteristic** of a typical **CMOS** NOT gate run off supply rails of +5 V and 0 V. (CMOS (pronounced 'sea moss') is the name given to one type of logic gate. Different types (such as TTL and E^{2}L) have different transfer characteristics. They will not be considered in this book.) The graph shows that when V_{IN} is below +2.5 V then V_{OUT} will be +5 V. Conversely, if V_{IN} is between +2.5 V and +5 V then V_{OUT} will be 0 V.

**When the input is high, the output is low;
when the input is low, the output is high.**

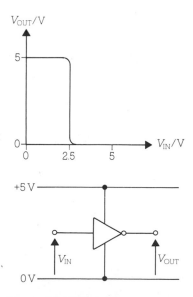

Figure 5.2 The transfer characteristic of a NOT gate

When we are dealing with logic gates, it is standard practice to use **1** (one) and **0** (nought) instead of **high** and **low**. For CMOS logic gates run off supply rails of V_{CC} and 0 V, a 1 represents any signal between V_{CC} and $\frac{1}{2}V_{CC}$ and a 0 represents any signal between $\frac{1}{2}V_{CC}$ and 0 V. V_{CC} can have any value between +3 V and +18 V.

Truth tables

A **truth table** is a very convenient way of summarising the behaviour of a logic gate. Here is the truth table of a NOT gate.

A	Q
0	1
1	0

The table has two columns. The A column refers to the input, the Q column to the output. Each row of the table tells you what the output is for a particular input. So the first row tells you that when A is 0 then Q is 1. Similarly, the second row tells you that when A is 1 then Q is 0.

For a NOT gate, Q is 1 only if A is **not** 1.

The AND gate

The circuit symbol for an **AND gate** is shown in figure 5.3. It has two inputs (marked A and B) and one output (marked Q). Its truth table is shown below.

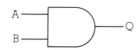

Figure 5.3 An AND gate

B	A	Q
0	0	0
0	1	0
1	0	0
1	1	1

For an AND gate, Q is 1 only if A **and** B are 1.

The OR gate

The circuit symbol for an **OR gate** is shown in figure 5.4. It too has two inputs (A and B) and a single output (Q). Here is its truth table.

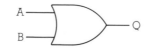

Figure 5.4 An OR gate

B	A	Q
0	0	0
0	1	1
1	0	1
1	1	1

For an OR gate, Q is 1 only if A **or** B are 1

Using logic gates

The **pinout** of the ICs which contain AND, OR and NOT gates are shown in figure 5.5. Each IC contains just one type of logic gate. So the 4069 IC contains six NOT gates which share power supply connections; V_{CC} is usually +5 V and GND is usually 0 V. Similarly, the 4081 IC contains four independent AND gates built onto the same small crystal of silicon.

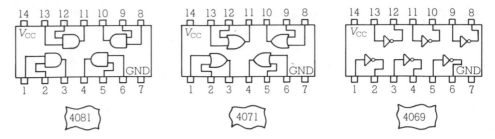

Figure 5.5 Some CMOS logic gate ICs

Inputs and outputs

Figure 5.6 shows a circuit which uses one AND gate and one OR gate. The inputs of CMOS logic gates require virtually no current to go in or out of them, so they do not upset the operation of voltage dividers.

CMOS gate outputs can only provide a current of at most 2 mA, so they need to be buffered if they are to drive more than just a voltmeter. So the circuit contains a couple of 555 ICs to allow the system to control a buzzer and an LED.

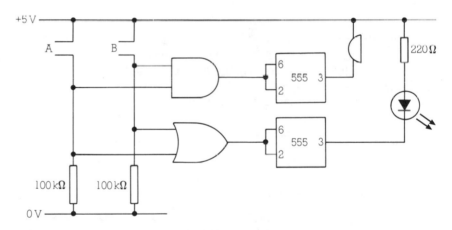

Figure 5.6 Logic gates combining signals from two sensors

Wetness indicator

A glance at the truth table of an AND gate will tell you that the buzzer will only make a sound when both of the contacts are wet. Similarly, because of its truth table, the OR gate ensures that the LED will glow if either, or both of the contacts are wet. The behaviour of the whole system is summarised in the table below.

A	B	LED	buzzer
dry	dry	off	off
dry	wet	on	off
wet	dry	on	off
wet	wet	on	on

Combining logic gates

When you connect logic gates to each other, you usually end up with a system which has a truth table which is not the same as the truth tables of the individual gates.

NAND gates

For example, two gates have been connected together in figure 5.7. The whole system has inputs A and B and an output Q. We can work out the truth table of that system by considering the truth tables of an AND gate and a NOT gate in turn.

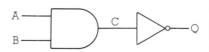

Figure 5.7 A NAND gate from an AND gate and a NOT gate

The first step is to label the outputs of all the gates and write down a blank truth table. So we call the output of the AND gate C (it could be anything which was not A, B or Q). Since the truth table has to contain all possible combinations of A and B it needs $2^2 = 4$ rows.

B	A	C	Q
0	0	?	?
0	1	?	?
1	0	?	?
1	1	?	?

The next step is to fill in the C column. C is the output of an AND gate whose inputs are A and B. So C can only be 1 when both A and B are 1. The last row of the table must therefore have C equal to 1; the other three rows must have C equal to 0.

At this stage the truth table looks like this.

B	A	C	Q
0	0	0	?
0	1	0	?
1	0	0	?
1	1	1	?

The final step is to write down the Q column. Q is the output of a NOT gate whose input is C. So Q is the opposite of C. Therefore the first three rows of the truth table have Q equal to 1, and the last row has Q equal to 0. The final truth table is shown below.

B	A	C	Q
0	0	0	1
0	1	0	1
1	0	0	1
1	1	1	0

The whole system has the truth table of a NAND gate. Its symbol is shown in figure 5.8.

For a NAND gate, Q is **not** 1 only if A **and** B are 1.

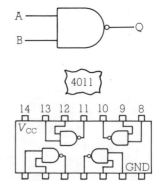

Figure 5.8 A NAND gate

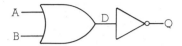

Figure 5.9 A NOR gate from an OR gate and a NOT gate

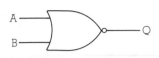

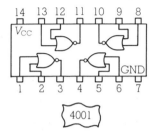

{4001}

Figure 5.10 A NOR gate

The NOR gate

A **NOR gate** can be made out of an OR gate and a NOT gate as shown in figure 5.9. Its circuit symbol is shown in figure 5.10. Its truth table can be worked out exactly as that for the NAND gate was. Here is the complete truth table. You can check it through for yourself.

B	A	D	Q
0	0	0	1
0	1	1	0
1	0	1	0
1	1	1	0

For a NOR gate, Q is **n**ot 1 only if A **or** B are 1

QUESTIONS

1 For each of the logic gates listed below write down
   a) its circuit symbol,
   b) its truth table,
   c) a sentence describing its behaviour.

                NOT AND OR NAND NOR

2 Explain what the terms 1 and 0 mean.

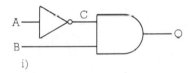

i)

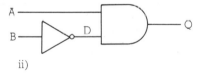

ii)

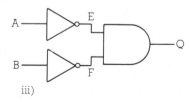

iii)

Figure 5.11 Question 3

3 Each of the three systems shown in figure 5.11 has a different truth table. In each case, only one of the rows will have Q equal to 1. Work out the truth table of each system.

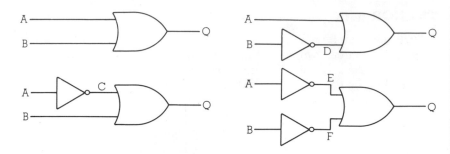

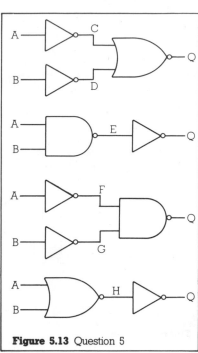

Figure 5.12 Question 4

4 Each of the four systems shown in figure 5.12 has only one row of their truth tables with Q equal to 0. Work out their truth tables.

5 The circuits of figure 5.13 contain NOT, NAND and NOR gates. Each system behaves like either an AND gate or an OR gate. Work out the truth table of each system and state what type of logic gate it behaves like.

6 The two circuits of figure 5.14 illustrate how you can build any logic gate from just AND and NOT gates or just OR and NOT gates. Work out the truth table for each system and state which type of logic gate it behaves like.

Figure 5.13 Question 5

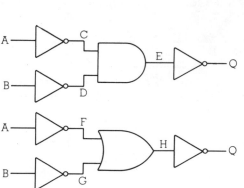

Figure 5.14 Question 6

7 The two systems shown in figure 5.15 have three inputs, labelled C, B and A. This means that their truth tables must have $2^3 = 8$ rows.
a)   Work out the truth table of each circuit.
b)   Describe the behaviour of each circuit in words.

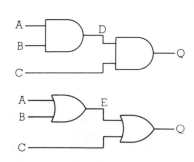

Figure 5.15 Question 7

6
Designing logic systems

This chapter is going to show you how to combine logic gates to make useful systems. An enormous variety of logic systems can be made by interconnecting NOT, AND and OR gates. This means that ICs containing the basic logic gates are made in enormous numbers. This makes them cheap to produce. So rather than build an IC to do a particular task, it is usually more cost-effective to put together a number of basic ICs instead.

The Exclusive–Or gate

The first system we are going to design must have the following behaviour.

**The output (Q) must only go high when
the inputs (A and B) are different.**

Systems with this behaviour are called **Exclusive-Or gates**. Their truth table is shown below.

B	A	Q
0	0	0
0	1	1
1	0	1
1	1	0

Two of the rows have Q equal to 1. So we will have to add two columns to the truth table, headed (at random) C and D. Each of these columns must contain 1 only once, as shown below.

B	A	C	D	Q
0	0	0	0	0
0	1	0	1	1
1	0	1	0	1
1	1	0	0	0

To understand why we have added these columns, look at the circuit of figure 6.1. It shows the final circuit for the system we are trying to design. The OR gate (marked 5) has Q as its output and C and D as its inputs.

Now we have to work out how to generate C and D from B and A. Let's start off with C.

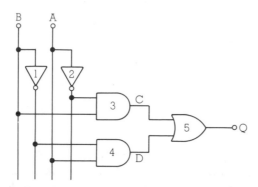

Figure 6.1 An Exclusive-Or gate

From the third line of the truth table you can see that C is only 1 when A is 0 and B is 1. In figure 6.1 C is generated by an AND gate which will only feed out 1 when both its inputs are 1. So a NOT gate has been placed between the input A and the AND gate, to convert 0 to 1.

A similar technique has been used to generate D. The NOT gate marked 1 changes the state of B before it gets to the AND gate marked 4. So D can only be 1 when B is 0 and A is 1, as required by the truth table.

The switched inverter

Our second example is a very useful component of electronic memory systems. It has two inputs (A and ON) and two outputs (X and Y). When ON is a 1, then X is the same as A and Y is the opposite of A. When ON is a 0, then both X and Y are 0.

As usual, the starting point for the design is a truth table.

ON	A	X	Y
0	0	0	0
0	1	0	0
1	0	0	1
1	1	1	0

Now we look carefully at the columns for X and Y to see if there are any familiar patterns. Sure enough, X looks like the output of an AND gate whose inputs are A and ON. That gate in the final system (which is shown in figure 6.2) has been labelled 1.

Y is 1 only once, so we should be able to generate it with an AND gate. From the truth table, Y is 1 when ON is 1 and A is 0. By inverting A with a NOT gate and combining it with ON via an AND gate we can generate Y.

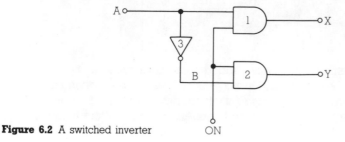

Figure 6.2 A switched inverter

Figure 6.2 has been drawn to emphasise the way in which ON controls X and Y via the two AND gates. When ON is a 0, both X and Y **have** to be 0, regardless of what A does. This use of AND gates to control the flow of information is widely used; figure 6.4 shows another example.

The keyboard encoder

It is common practice for people to feed information into an electronic system with the help of a keyboard. For example, you have probably given instructions to a computer via a keyboard. The computer had to work out which keys you were pressing. An **encoder** is a system which can tell other systems, using a specific **code**, which key is being pressed at any instant.

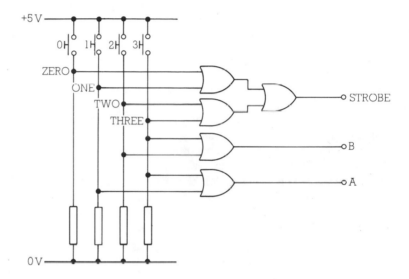

Figure 6.3 A keyboard encoder

Figure 6.3 is the circuit diagram of a very simple encoder. On the left there are four keys labelled 0, 1, 2 and 3. Each time that one key is pressed, the outputs (STROBE, B and A) inform some other system (not drawn) which of the keys is being pressed.

The STROBE output only goes high when a key is being pressed. The other two outputs use a **binary code** to say which key is being pressed. The table below shows which **binary word BA** is produced for each key.

Key pressed	Binary word BA
0	00
1	01
2	10
3	11

Having stated exactly what the outputs have to do when a key is pressed, we can draw up a truth table for the whole system. The four inputs to the logic gates from the switches have been called ZERO, ONE, TWO and THREE. Each input will only be a 1 when its switch is being pressed.

ZERO	ONE	TWO	THREE	STROBE	B	A
0	0	0	0	0	X	X
1	0	0	0	1	0	0
0	1	0	0	1	0	1
0	0	1	0	1	1	0
0	0	0	1	1	1	1

The first line of the table refers to when none of the keys are being pressed. STROBE must be 0, but we don't care what B and A are. So we have entered X for them to signify 'don't care'. The other four lines correspond to just one of the keys being pressed. We are not interested in what happens when two or more keys are pressed.

STROBE is 1 when any of the inputs are 1. So it can be generated by a four input OR gate or its equivalent made from three two input OR gates.

B is 1 when TWO or THREE are 1, so it can be generated by an OR gate. Similarly, A can be generated from ONE and THREE via another OR gate.

The binary decoder

The **binary encoder** of the last example looks at a number of inputs and feeds out a **binary code** which says which of the inputs is high. A binary decoder does exactly the opposite. It reads in a binary code, unscrambles it and raises one of a number of outputs up to 1.

We want to design a **two bit binary decoder**. Its input will be a binary word BA which can take one of four values (00, 01, 10 and 11). So it will have four outputs which we shall call ZERO, ONE, TWO and THREE. We shall also need a STROBE input to tell the decoder when it is being fed a binary word. So when STROBE is 0, all four output lines must be 0.

Here is the truth table of the decoder.

B	A	STROBE	THREE	TWO	ONE	ZERO
X	X	0	0	0	0	0
0	0	1	0	0	0	1
0	1	1	0	0	1	0
1	0	1	0	1	0	0
1	1	1	1	0	0	0

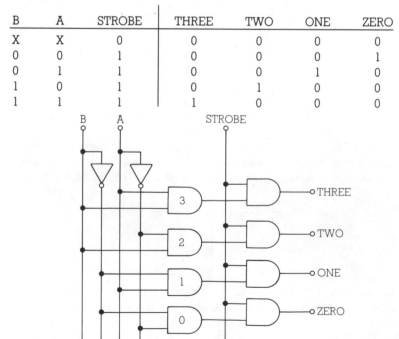

Figure 6.4 A binary decoder

The final circuit is shown in figure 6.4. The binary word BA is fed in top left and the four outputs are on the right. STROBE is used to force the outputs to 0 via four AND gates. The other AND gates (marked 0, 1 , 2 and 3) each generate a 1 for a different binary word. For example, when BA is 10, only the output of gate 2 will be 1. If STROBE is 1 as well, then TWO will be a 1 as required by the truth table.

QUESTIONS

1   Design a logic system which has one output (Q) and two inputs (A and B). Q must only be 1 when A is 1 and B is 0, otherwise Q must be 0.

2   A logic system has two inputs (C and D) and a single output (R). The output is 1 only when the two inputs have the same state as each other. Design a suitable logic system.

3   You have to design an encoder circuit which is similar to that of figure 6.3. The four keys are labelled F, B, L and R. The three outputs of the encoder are labelled GO, Y and Z. The whole system has to behave as shown in the table below.

Key pressed	GO	Y	Z
none	0	X	X
F	1	1	0
B	1	1	1
L	1	0	1
R	1	0	0

Draw a circuit diagram for the whole system, including the keys.

4   Design a decoder circuit which obeys the following truth table. The two inputs are B and A. The three outputs are R, S and T.

B	A	R	S	T
0	0	0	0	0
0	1	1	0	0
1	0	1	1	0
1	1	1	1	1

7
Doing it with NAND gates

So far you have been designing logic systems with the help of three types of gate; AND gates, OR gates and NOT gates. This can lead to considerable waste when circuits are assembled, because you may only need to use some of the gates in an IC. For instance, you may need one OR gate in your circuit but you still have to plug four of them into your breadboard. You can avoid this waste by only using one type of gate in your circuit, the NAND gate. It turns out that it is always possible to build a logic system out of just NAND gates. This is because NAND gates can be used to make NOT, AND and OR gates if you connect them together correctly.

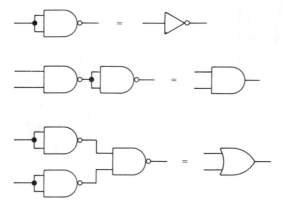

Figure 7.1 How NAND gates can be used to make basic logic gates

Combining NAND gates

Figure 7.1 shows how NAND gates have to be connected to make each of the basic logic gates.

Consider the top circuit, the NOT gate. The two inputs of the NAND gate have been connected together to make a single input. The behaviour of the resulting system can be worked out by referring to the NAND gate truth table.

B	A	Q
0	0	1
0	1	1
1	0	1
1	1	0

The first line of the truth table tells you that when both inputs of the NAND gate are 0 the output will be 1. Similarly, the last line tells you that when both inputs are 1 the output is 0. So the whole system behaves like a NOT gate.

The questions at the end of the chapter will invite you to work out the truth tables for the other two circuits in figure 7.1. The centre one has the truth table of an AND gate and the bottom one has the truth table of an OR gate.

Condensing a circuit

At the top of figure 7.2 there is a logic system made from an AND gate, an OR gate and a NOT gate. You are going to be shown how to design a system with the same truth table, but made from only NAND gates.

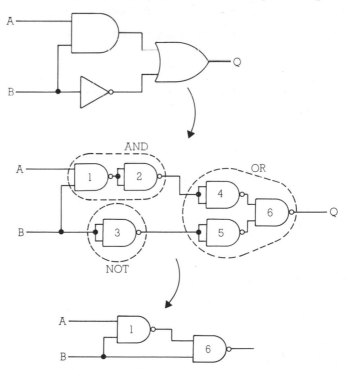

Figure 7.2 Converting a circuit with a mix of gates into one which only uses NAND gates

The first step is to redraw the circuit, replacing each gate with its NAND gate equivalent. This is shown in the centre of figure 7.2.

The second step is to look for adjacent pairs of NOT gates. There are two such pairs in our example. Gates 2 and 4 form a pair of NOT gates, as do gates 3 and 5. Now whatever signal you feed into a pair of NOT gates in series, you will get the same signal out. So a 1 fed into gate 2 causes gate 4 to also feed out a 1. Therefore pairs of NOT gates in series can be replaced by a length of wire without altering the overall behaviour of the system.

The final circuit, with gates 2, 3, 4 and 5 removed, is shown at the bottom of figure 7.2. The original circuit contained three gates, whereas the final one only contains two. Furthermore, the original circuit would have required three ICs, with most of the gates in those ICs being left unused. The final circuit could be assembled with just one IC. These are the typical benefits which are gained by condensing a circuit containing NOT, AND and OR gates into a NAND-only one.

QUESTIONS

1 Copy and complete the following statements.
   When both inputs of a NAND gate are 0, the output is . . . .
   When both inputs of a NAND gate are 1, the output is . . . .
   So a NAND gate which has both of its inputs connected together behaves like a . . . gate.

2 For each of the circuits shown in figure 7.3, draw up a truth table. Each table should contain columns for the outputs of all the gates in the circuit. State which type of gate each system behaves like (if any).

3 Redraw each of the following circuits, using NAND gates only.
   a) Figure 6.2.
   b) Figure 6.1.
   c) Figure 5.15.
   d) Figure 5.12.

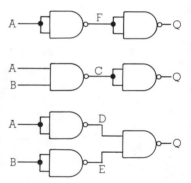

Figure 7.3 Question 2

Revision questions for Section A

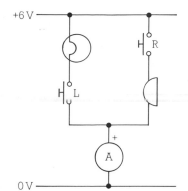

Figure A.1 Question 1

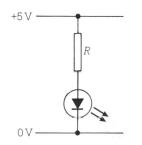

Figure A.2 Question 2

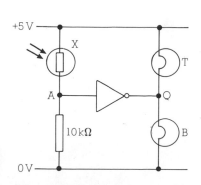

Figure A.3 Question 3

1 Figure A.1 shows a light bulb and a buzzer connected to a 6 V power supply. The bulb is rated at 6 V, 80 mA and the buzzer at 6 V, 20 mA.
   a) Are the bulb and buzzer connected in series or parallel with the power supply?
   b) Which of the switches must be pressed to make
      i)   the buzzer make a noise,
      ii) the bulb give out light?
   c) Complete the table shown below.

Switch L	Switch R	Ammeter reading
open	open	?
open	closed	?
closed	open	?
closed	closed	?

   d) Which of the components has the greater power?

2 Figure A.2 shows an LED in series with a resistor R. The LED is rated at 2.0 V, 15 mA. The value of R is such that a current of 15 mA flows through the LED.
   a) Draw a circuit diagram to show how you would connect a voltmeter to measure the voltage drop across the resistor. What would you expect the voltmeter reading to be?
   b) How much current goes through R?
   c) Calculate the resistance of R. Choose the nearest preferred resistance value and work out its colour code. Select a suitable power rating for the resistor.

   The resistor is now replaced with one that has a resistance of 470 Ω and a power rating of 125 mW.
   d) Calculate how much current will flow through the LED now. Will it be brighter or dimmer than before?
   e) Calculate the rate at which heat is being generated in the resistor. Is it in danger of overheating?
   f) Draw a circuit diagram to show how you would connect an ammeter to the circuit to measure how much current goes through the LED.
   g) In what way will the ammeter reading change if you insert the LED the other way round in the circuit?

3 The NOT gate of figure A.3 has a buffered output which can handle currents of up to 100 mA.
   a) Describe the behaviour of a NOT gate.
   b) What is the name of the component marked X? Describe its behaviour.
   c) What do you have to do to X to make its resistance much less than 10 kΩ? Roughly what is the voltage at A when you do this?
   d) What is the state of the light bulbs when A is
      i)   low,
      ii) high?
   e) Describe the overall behaviour of the system.

4 You have to design a rain detector for domestic use. Whenever rain falls on a suitable sensor placed outside the house, a light bulb and a buzzer must come on inside the house. This will enable the people in the house to get the washing inside before it gets too wet. You have to use a 12 V, 1 A bulb and a 12 V, 50 mA buzzer.
  a) Draw a block diagram of the system. Explain the function of each of the blocks.
  b) Draw a circuit diagram of the system. Explain how your system works.

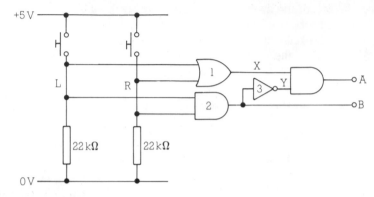

Figure A.4 Question 5

5 This question is about the circuit shown in figure A.4.
  a) For each of the gates which are numbered,
     i)   state their names,
     ii) describe their behaviour in words,
     iii) draw up their truth tables.

  When the switch on the left is pressed, the input L rises from 0 to 1.
  b) Explain what is meant by the terms 0 and 1.
  c) Complete the truth table shown.
  d) Draw diagrams to show how the following gates may be made from only NAND gates.
     i)   a NOT gate
     ii) an AND gate
     iii) an OR gate
  e) Draw an equivalent circuit for figure A.4 which only uses NAND gates.

L	R	X	Y	B
0	0	?	?	?
0	1	?	?	?
1	0	?	?	?
1	1	?	?	?

6 A plant protection system has four sets of contacts rammed into the soil of four flower pots. If any one of those flower pots runs dry, a buzzer must make a sound.
  a) Draw a block diagram for the system.
  b) Draw a circuit diagram for the system.
  c) Explain how your system works.

7 The invention of the integrated circuit is responsible for the rapid growth of the electronics industry in the last twenty years.
  a) What is the difference between an integrated circuit and one made of discrete components?
  b) In what ways are systems built of integrated circuits better than ones made of discrete components?

Telling the time

Using an oscilloscope to display the output waveform of a signal generator. *(Scopex Electronics)*

8
Capacitors

The circuit of figure 8.1 is an example of a system which has a **static** output. As soon as the LDR is covered over, the bulb will light up. Similarly, if the LDR is subsequently uncovered the bulb goes off immediately.

The addition of a **capacitor** to the circuit converts it into a **dynamic** system. This is shown in figure 8.2. The capacitor prevents A from going low immediately when the LDR is covered. So there is a time delay between the LDR being plunged into darkness and the bulb coming on.

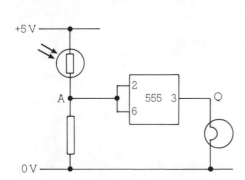

Figure 8.1 A static system

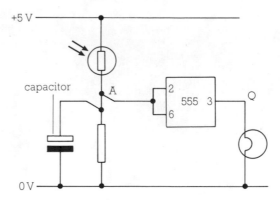

Figure 8.2 A dynamic system

Capacitor behaviour

A capacitor is basically a rechargeable battery. It acts like a short-lived power supply which can be quickly revitalised.

Capacitors come in many shapes and sizes, and are measured in units called **farads**. The **microfarad** or **µF** is a more convenient unit for electronics. There are 1 000 000 microfarads in 1 farad. Figure 8.3 contains a 1000 µF capacitor. It has two **plates** which are separated by a thin insulating layer.

Charging

In figure 8.3 the top plate of the capacitor is connected to the +5 V supply rail via a switch. As soon as that connection is made, a brief pulse of current appears to flow though the capacitor. This **charges up** the capacitor. Once it is charged up, no more current goes through it.

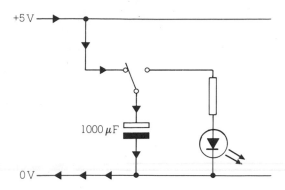

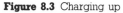

Figure 8.3 Charging up

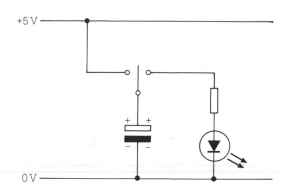

Figure 8.4 Isolated

Isolation

When the capacitor is **isolated** (figure 8.4), it stays charged. The top plate is positively charged, the bottom one negatively charged. This is shown by the + and − next to the capacitor plates. **An ideal capacitor which is isolated will stay charged for ever.** In reality the charge will slowly **leak** between the plates, so that the capacitor gradually returns to its initial state.

Discharging

Figure 8.5 shows the capacitor being **discharged**. The switch contacts have been arranged so that there is a complete circuit between the two plates. So a current goes from the positively charged plate to the negatively charged one, lighting up the LED on the way. After a while the capacitor will become completely discharged, and the current will stop. So although the LED initially glows brightly, it slowly gets dimmer and dimmer as the capacitor discharges.

**Large capacitors store more charge than small ones.
So they can keep the current going for longer.**

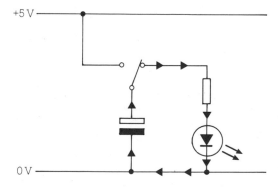

Figure 8.5 Discharging

Electrolytics

The type of capacitor used in figure 8.5 is known as an **electrolytic** capacitor. This refers to the way that it is constructed and the physics behind its operation. They must always be charged up such that the white plate is at a higher voltage than the black one. If you try charging up an electrolytic capacitor the wrong way round, it will stop being a capacitor and there is a risk that it will blow up.

Ratings

All electrolytic capacitors have a **voltage rating**. This is usually printed on the body (or **can**) of the capacitor. It is the maximum voltage that you can put across the plates of the capacitor without damaging it. If you try to charge up a capacitor from a power supply whose voltage is greater than the voltage rating, there is a good chance that the capacitor will explode.

Voltage–time characteristics

Figure 8.6 shows a capacitor in action. The LED stays on for about two minutes after the switch is briefly pressed and released. This how it works.

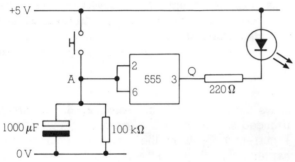

Figure 8.6 Using a capacitor to create a time delay

When the switch is closed the capacitor is quickly charged up from the +5 V power supply. So A goes high immediately, Q goes low immediately and the LED comes on.

As soon as the switch is released, the capacitor starts to discharge itself through the 100 kΩ resistor. (Only a negligible amount of current has to go into the 555 inputs.) So the voltage at A starts to drop as shown in the graph of figure 8.7.

While the voltage at A is high Q will remain low, lighting up the LED.

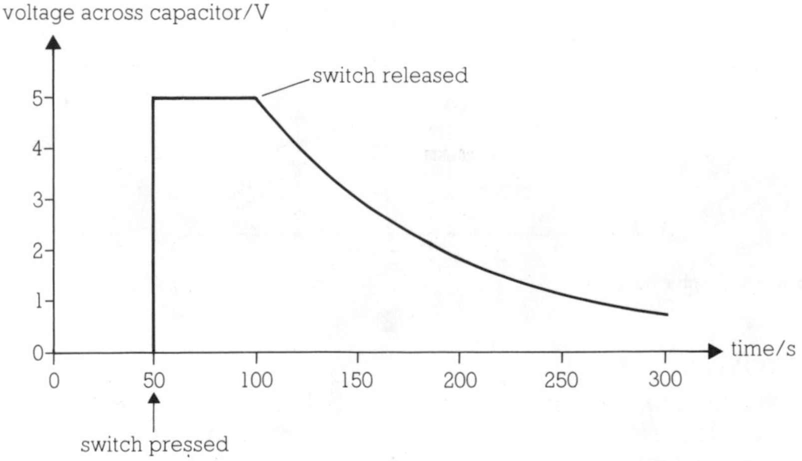

voltage across capacitor/V

Figure 8.7 How the voltage across the capacitor changes with time

After a couple of minutes A goes low. Q goes high immediately, turning off the LED. The capacitor will continue to discharge itself through the resistor. The voltage at A will drop lower and lower as time goes on.

The time constant

The length of time for which the LED of figure 8.6 glows after the switch is released depends on the resistor and capacitor used in the discharge circuit.

A large capacitor will hold a lot of charge, so it will keep the current going through the resistor for a long time. On the other hand, if the resistor is small a lot of current will flow through it, rapidly discharging the capacitor. So both the resistor R and capacitor C have to be carefully chosen to obtain a certain discharge time.

For the circuit of figure 8.6, the LED remains lit for a time T after the switch is released; **(1000 milliseconds = 1 second)**.

$$\boldsymbol{T = RC}$$ T is in milliseconds (ms)
R is in kilohms (kΩ)
C is in microfarads (μF)

The product RC is called the **time constant** of the circuit. Its value fixes how rapidly the capacitor C discharges itself through the resistor R.

**The time delay built into a circuit will be
proportional to its time constant.**

So how long does the LED remain lit after the switch of figure 8.6 is released?

$T = RC$ $T = ?$ therefore $T = 100 \times 1000$
$R = 100$ kΩ therefore $T = 100\ 000$ ms
$C = 1000\ \mu$F   therefore $T = 100$ s

The LED stays lit for 100 s.

Changing the delay

The output of the AND gate shown in figure 8.8 remains high for 15 s after the switch is released. What will happen to the time delay if we change the capacitor or the resistor?

If we double the time constant, we double the time delay. So replacing the 100 μF capacitor with a 200 μF one will increase the time delay to

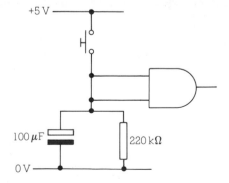

Figure 8.8 Fifteen second time delay

30 s. Alternatively, replacing the 220 kΩ resistor with a 440 kΩ one will also increase the time delay to 30 s. If we double both the resistor and the capacitor values, the time delay will be quadrupled i.e. become 60 s.

The change of time delay is summarised in the table below.

resistor	capacitor	time delay
R	C	T
R	$2C$	$2T$
$2R$	C	$2T$
$2R$	$2C$	$4T$

You may have noticed that the time delay is not equal to the time constant as it was for the 555 circuit of figure 8.6. In fact, for a CMOS logic gate the time delay is $0.7RC$.

Parallel capacitors

If you put two identical capacitors in parallel with each other, as shown in figure 8.9, you end up with a system which has twice the capacitance of each capacitor.

So two 1000 μF capacitors in parallel behave like a single 2000 μF capacitor. Similarly a 1000 μF capacitor in parallel with a 470 μF capacitor behaves like a 1470 μF capacitor. Since capacitors are not made in the same wide range of values that resistors are, you quite often have to use such parallel connections to obtain exactly the capacitance that you need.

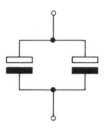

Figure 8.9 Capacitors in parallel

Delaying logic signals

Since capacitors are short-lived power supplies they can be used to hold part of a circuit high for a time. (As in figure 8.8). They can also be used to **delay** the change of a logic signal in a circuit. So if an input rises from 0 to 1 there is a time delay before the output changes.

The circuit of figure 8.10 shows how this can be done. If A changes state (1 to 0 or 0 to 1), then there is a delay of T before Q changes state. The time delay is caused by the resistor R and the capacitor C connecting the output of the first NOT gate to the input of the second one.

The graphs of figure 8.10 show what happens when A changes state. The voltage at point B can only change slowly because the resistor restricts the charging and discharging currents. So the voltage at B climbs from 0 V to +2.5 V in $0.7RC$ ms when A falls from 1 to 0. Similarly, B drops from +5 V to +2.5 V in $0.7RC$ ms each time that A rises from 0 to 1.

When A changes state Q changes $0.7RC$ ms later.

Note how the time delay depends on the time constant (RC) of the circuit. It is useful to remember that a capacitor will be almost completely charged (or discharged) in about three time constants. So the circuit of figure 8.10 will only work properly if the changes of A are at least $3RC$ ms apart.

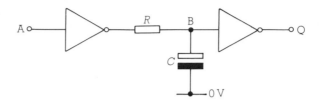

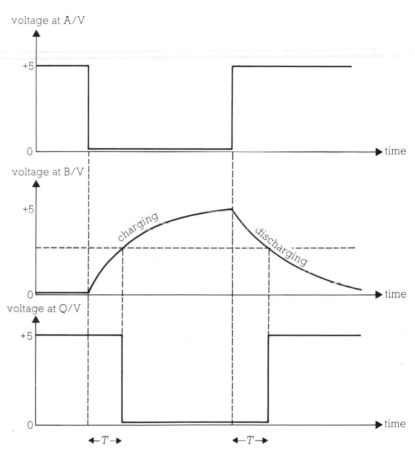

Figure 8.10 Delaying logic signals

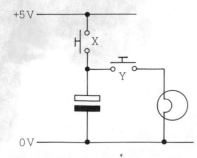

Figure 8.11 Question 1

QUESTIONS

1 This question is about the circuit shown in figure 8.11.
   a) Which switch must be pressed to
       i)   charge up the capacitor,
       ii) discharge the capacitor?
   b) Y is pressed immediately after the capacitor has been charged up. What happens to the light bulb?
   c) A second identical capacitor is put in parallel with the one already in the circuit. What effect does this have on the bulb as the capacitors are discharged?

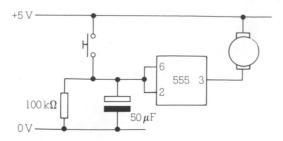

Figure 8.12 Question 2

2 The circuit of figure 8.12 contains, among other things, a push switch and a motor.
   a) Describe what happens to the motor when the push switch is
     i) pressed,
     ii) released.
   b) The motor spins for 5 s when the switch is released with the values of resistor and capacitor shown. For how long will it spin if
     i) the resistor becomes 200 kΩ,
     ii) the capacitor becomes 25 μF.
     iii) the values of both capacitor and resistor are doubled?
   c) Suggest values for the resistor and capacitor which will allow the motor to spin for 30 s when the switch is released.
   d) Sketch a graph to show how the voltage at the input of the 555 changes with time when the switch is released. Mark the axes carefully.

3 The circuit of figure 8.13 contains a **tilt switch** as an input transducer. It is a pair of contacts in a glass bulb with a blob of mercury. When the tilt switch is vertical (see figure 8.13) the mercury connects the

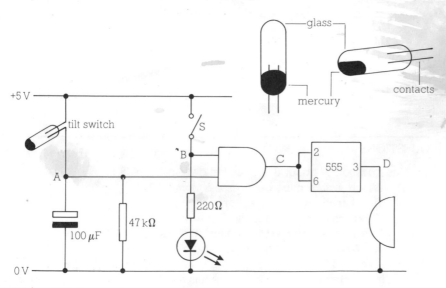

Figure 8.13 Question 3

contacts, so that there is very little resistance between them. But when the tilt switch is not vertical, there is a very high resistance between the contacts.

a) What is the state of point B when the switch S is
    i) open,
    ii) closed?

b) What happens to the voltage at point A when the tilt switch is
    i) upright,
    ii) on its side?

c) Copy and complete the following statements.
    When the tilt switch is upright and the switch is closed, C is a . . . . and the buzzer . . . . . If the switch S is opened the LED . . . . and the buzzer . . . . . Alternatively, if the tilt switch is put on its side there is a delay of about . . . . seconds before A is a . . . . and the buzzer . . . . .

d) The system of figure 8.13 was designed to be run off a battery so that it could be fastened to the clothing of an elderly infirm person. Explain how the circuit could be used to send out an alarm signal when that person fell over and failed to get up straight away. What is the purpose of the LED in the circuit?

4 This question is about the circuit shown in figure 8.14. Copy and complete the following statements.

    E will go to 1 immediately when A is . . . . and B is . . . . . When this happens, D will go to . . . . after a time delay. The length of the delay is fixed by the . . . . of the circuit which is . . x . . So when both A and B are 1 there is a delay before Q is . . . . . The delay could be doubled by . . . . the size of R. It could be halved by . . . . the size of C.

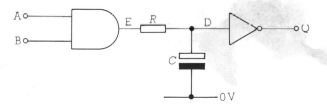

Figure 8.14 Question 4

9

Relaxation oscillators

An **oscillator** is an electronic system which cannot settle into just one state. It changes state at regular intervals.

For example, the LED in the circuit of figure 9.1 flashes on and off continually. It comes on for one second. Then it goes off for one second. Then it comes on for second. And so on, until the power supply is switched off. The circuit is one example of a **relaxation oscillator**. It is a **Schmitt trigger NOT gate** whose output is fed back into its input via a time delay.

Schmitt trigger NOT gates

The circuit symbol for a **Schmitt trigger NOT gate** is shown in figure 9.2. It obeys the normal truth table of a NOT gate.

A	Q
0	1
1	0

The NOT gates in a 4069 IC which is being run off +5 V and 0 V treat any signal above +2.5 V as a 1 and any signal below +2.5 V as a 0. Schmitt trigger NOT gates, however, have transfer characteristics which depend on their output states. The state of Q fixes the range of input signals which will be recognised as a 1 or a 0.

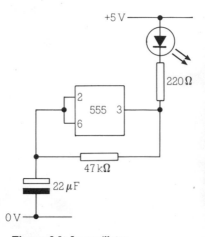

Figure 9.1 An oscillator

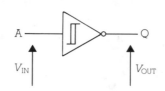

Figure 9.2 A Schmitt trigger NOT gate

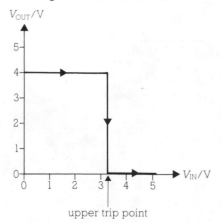

Figure 9.3 Transfer characteristic when the output is initially high

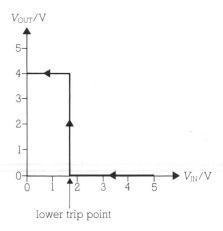

lower trip point

Figure 9.4 Transfer characterisitic when the output is initially low

For example, the graph of figure 9.3 shows the transfer characteristic of the NOT gate in a 555 IC when A is raised from 0 V to +5 V. When A gets to +3.3 V (the **upper trip point**) it ceases to be 0 and becomes 1.

Figure 9.4 shows what happens if A starts at +5 V and is lowered to 0 V. When A gets to +1.7 V (the **lower trip point**) it ceases to be 1 and becomes 0.

This behaviour of the Schmitt trigger NOT gate in a 555 IC is summarised in the table below.

Q	1	0
1	>3.3 V	<3.3 V
0	>1.7 V	<1.7 V

Oscillation

If you connect the output of a Schmitt trigger NOT gate to its input with a time delay, then it will oscillate.

Consider the circuit of figure 9.5. When the power supply is switched on, the capacitor C will be uncharged. So V_{IN} will be 0 V and the system will be at point U on the graph.

The capacitor will charge up from V_{OUT} via the resistor R. So the system moves gradually from U to V on the graph.

When V is reached, V_{OUT} goes low. So the system goes from V to W on the graph. This is because V_{IN} has reached the upper trip point. As V_{OUT}

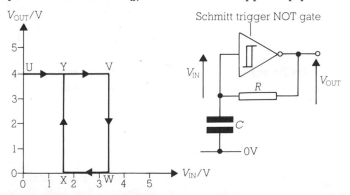

Figure 9.5 Making a Schmitt trigger NOT gate oscillate

is now 0 V, C proceeds to discharge itself through R, making the system gradually move from W to X on the graph.

At X, V_{IN} reaches the lower trip point. So the system moves rapidly from X to Y on the graph. Then gradually from Y to V, rapidly from V to W, gradually from W to X and so on.

**The oscillator takes about 2RC ms
to go through each cycle of oscillation.**

Waveforms

The standard way of displaying the output of an oscillator is with a voltage-time graph known as a **waveform**.

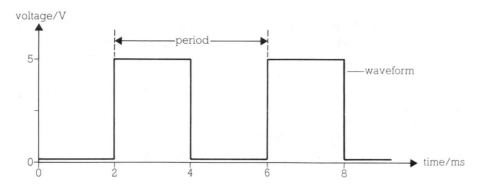

Figure 9.6 A square wave

Figure 9.6 shows a typical waveform for a relaxation oscillator. This type of waveform is usually known as a **square wave**, although it doesn't really look square at all. The waveform shows that the output alternates between being high (1) and low (0), spending an equal amount of time in each state.

The output of the oscillator consists of a pattern which is repeated time after time. The pattern is a 1 followed by a 0; it is called a **cycle**. The **period** of the waveform is the time taken for one cycle of oscillation. For the waveform of figure 9.6, the period is 4 ms.

Frequency

It is standard practice in electronics to quote a value for the **frequency** of an oscillator when you wish to say how fast it is oscillating.

**The frequency of a waveform is the number of
cycles it goes through in one second.**

The frequency can be calculated from the period with this rule.

$$f = \frac{1}{T}$$ f is the frequency in kilohertz.
T is the period in milliseconds.

Although the standard unit for frequency is the **hertz (Hz)**, it is often more convenient in electronics to use **kilohertz (kHz); 1000 Hz = 1 kHz**. So what is the frequency of the waveform shown in figure 9.6?

$$f = \frac{1}{T} \qquad \begin{matrix} f = ? \\ T = 4 \text{ ms} \end{matrix} \qquad \text{therefore } f = \frac{1}{4}$$

$$\text{therefore } f = 0.25 \text{ kHz}$$
$$\text{therefore } f = 250 \text{ Hz}$$

So the waveform goes through 250 cycles of oscillation in one second.

The oscilloscope

Oscillators which contain large capacitors oscillate slowly. They have long periods. So it is quite easy to study their outputs with the help of an LED or a voltmeter. But if you use small capacitors the changes become too rapid for you to be able to follow them easily. Once the frequency is above 25 Hz, the LED will not appear to flicker any more and will seem to be on all of the time. So if you want to study the waveform of a fast oscillator you need to use an instrument. That instrument is an **oscilloscope**.

Screen traces

Figure 9.7 shows the front view of a typical **cathode ray oscilloscope** or **CRO**. It has many control knobs, but only one screen. So we shall explain what the knobs do after you have found out what the screen does!

The screen displays the waveform of the signal being fed into the input.

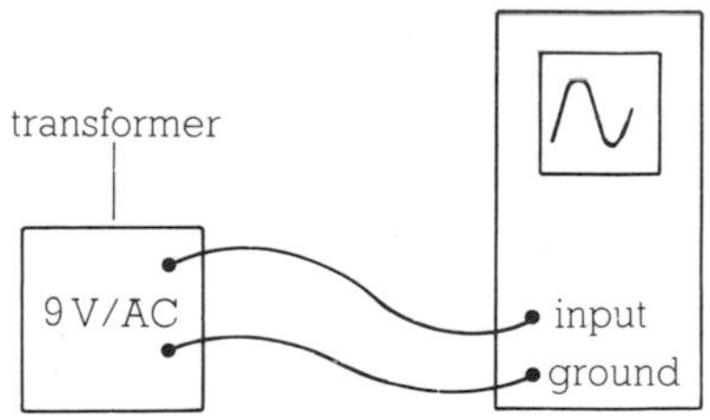

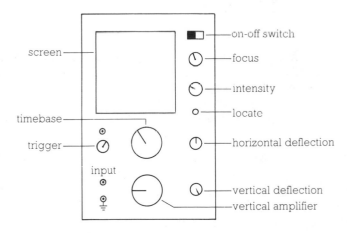

Figure 9.8 Connecting a signal source to a CRO

Figure 9.7 An oscilloscope

For example, suppose that the CRO was connected to a 9 V transformer as shown in figure 9.8. One of the input terminals of the CRO is held at 0 V (the one marked GROUND). The other one (marked INPUT) samples the voltage of the transformer output.

Figure 9.9 shows what appears on the CRO screen. It is called a **trace**, and it shows how the voltage of the input terminal is changing with time. The trace is a voltage-time graph, with time on the horizontal axis. The type of waveform shown in figure 9.9 is called a **sine wave**.

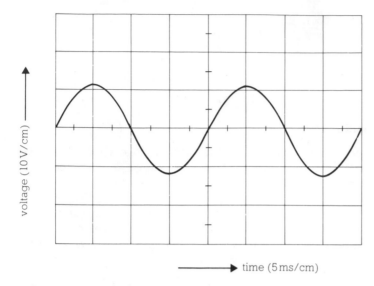

time (5 ms/cm)

Figure 9.9 A waveform displayed on a CRO screen

Measuring period

The screen is divided into 1 cm squares. These allow you to measure the period of the waveform. This is because the time-scale of the trace can be set with the **timebase control** (see figure 9.7). For figure 9.9 the timebase knob was set at 5 ms/cm. So each centimetre in the horizontal direction represents five milliseconds. As each cycle of the waveform is 4 cm long, the period must be $4 \times 5 = 20$ ms.

Measuring amplitude

The vertical scale of the trace is set by the **vertical amplifier**. For figure 9.9, the vertical amplifier knob was set at 10 V/cm. So each centimetre in the vertical direction represents 10 V, with the centre of the screen being 0 V. The peak of the waveform is about 1.2 cm above the central line on the screen, so its **peak value** is $1.2 \times 10 = 12$ V. The waveform has an average value of 0 V and an **amplitude** of 12 V.

Describing waveforms

Figure 9.10 is another example of a CRO trace. The vertical amplifier was set at 2 V/cm, so the waveform alternates between 0 V and +5 V. One

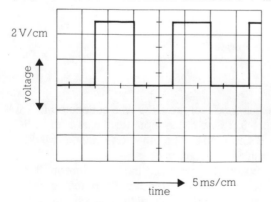

Figure 9.10 CRO trace of a 67 Hz square wave

cycle of oscillation of the square wave took 3 cm. As the timebase was set at 5 ms/cm, this means that the period of the waveform was 15 ms. Its frequency was 1 ÷ 15 = 0.067 kHz or 67 Hz.

Other controls

A number of control knobs are provided on the front of a CRO. Two of them (the timebase and vertical amplifier knobs) you know about already. They allow you to adjust the voltage- and time-scales of the screen.

Two knobs control the thickness and brightness of the trace. The **intensity** knob allows you to adjust the brightness of the trace. The **focus** knob is used to adjust the thickness of the trace. If the trace is too bright it may be difficult to focus down to a thin line.

The **vertical deflection** knob allows you to set the trace to 0 V correctly. By rotating it you move the trace up or down on the screen. Figure 9.11 shows what the trace should look like if no signal (i.e. 0 V) is fed into the CRO inputs. Similarly, the **horizontal deflection** knob allows you to move the trace left and right on the screen. It is used to centre the trace on the screen.

Finally, the **trigger** knob is the one which you adjust to get a steady trace on the screen. Its use is usually a matter of trial and error; you rotate it carefully until the trace stops moving across the screen.

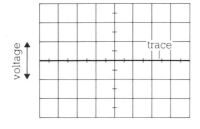

Figure 9.11 CRO trace for no input signal

The 555 oscillator

If you connect a resistor and a capacitor to a 555 IC you have a very useful and versatile oscillator. They are used all over the place in the rest of this book. That basic oscillator was shown in figure 9.1. A more refined version is shown in figure 9.12.

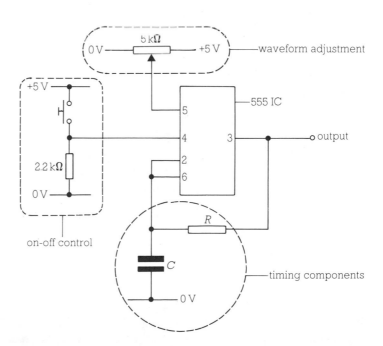

Figure 9.12 A 555 relaxation oscillator

The resistor R and capacitor C are the **timing components**. They fix the frequency of oscillation, which is given by this approximate rule.

$$f \simeq \frac{1}{2\,RC}$$

f is measured in kilohertz,
R is measured in kilohms,
C is measured in microfarads.

On and off

Pin 4 can be used to switch the oscillator on and off. If pin 4 is low, then pin 3 (the 555's output) **has** to be low. So the system of figure 9.12 will only oscillate when the switch is pressed. This **control pin** means that a 555 oscillator can be switched on and off by other electronic systems. If no connections are made to pin 4 it thinks that it is high and will allow the 555 to oscillate.

Waveform shape

If you leave pin 5 (the **waveform control**) alone the output of the oscillator will not be a square wave. It will spend longer at 1 than it does at 0. Quite often, this doesn't matter at all. You can however, adjust the squareness of the waveform with a 5 kΩ potentiometer as shown in figure 9.12.

Figure 9.13 is a typical CRO trace obtained of the waveform from a 555 oscillator. The voltage of pin 5 has been adjusted so that the waveform is square. The period is 2 ms, corresponding to R equal to 10 kΩ and C equal to 0.1 μF.

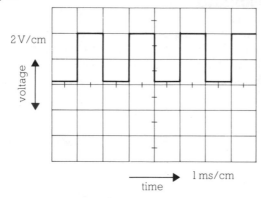

Figure 9.13 Typical CRO trace of 555 oscillator output

Listening to waveforms

An oscilloscope is not the only device which you can use to monitor the frequency of an oscillator. You can use a **loudspeaker** to generate sound waves and listen to their pitch instead.

Figure 9.14 shows how you can do it. On the left is a 555 oscillator, with an LDR as one of the timing components. The other timing component is a 100 nanofarad (**nF**) capacitor; (**1000 nF = 1 μF**). Both of its plates are drawn black. This means that it is unpolarised and can be connected either way round in the circuit. As a general rule, capacitors greater than 1 μF have to be polarised.

The output of the oscillator is fed into a second 555 IC which is wired up as a buffer. The output of that buffer drives the speaker. **The buffer is necessary because speakers require a lot of current to produce a reasonable amount of noise**. If the speaker had been connected directly

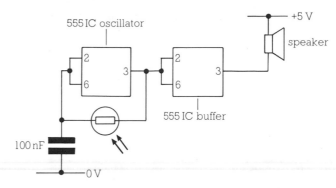

Figure 9.14 A light-to-sound converter

to the output of the oscillator it would have drawn enough current to stop the oscillator working properly.

It should be obvious that the frequency of the waveform being fed into the speaker will depend on the amount of light hitting the LDR. The frequency will be high when there is lots of light, giving a high pitched note from the speaker. The frequency (and hence the pitch of the note produced) will be low when the LDR is in the dark.

The system is a light-to-sound converter.

QUESTIONS

1   Figure 9.15 shows a number of CRO traces. Work out the following quantities for each of the waveforms.
   a) The maximum voltage.   c) The period (in ms).
   b) The minimum voltage.   d) The frequency (in kHz and Hz).

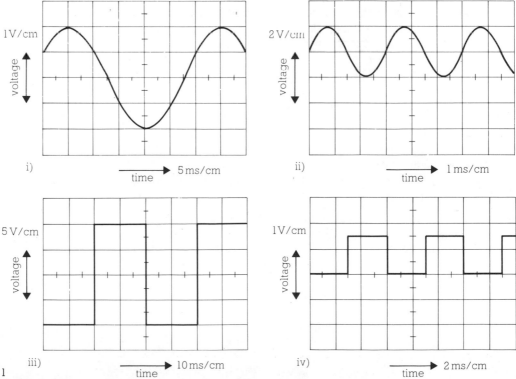

Figure 9.15 Question 1

2 The frequency of a 555 oscillator (in kHz) is roughly equal to $1/2 RC$ if R is in $k\Omega$ and C is in μF. R and C are the timing components. If you have to use a $0.1 \mu F$ capacitor, work out the resistance required to make the system oscillate with a frequency of

a) 10 kHz,
b) 2 kHz,
c) 500 Hz.

3 This question will show you how block diagrams can help you to analyse the behaviour of a system. The circuit diagram is shown in figure 9.16. When the switch is pressed and released, the bulbs are supposed to flash on and off for a certain length of time.

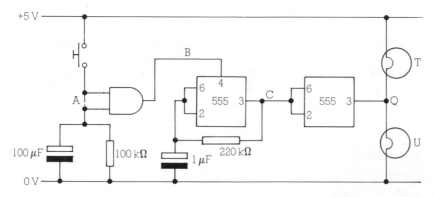

Figure 9.16 Question 3

a) Draw a block diagram of the circuit, using the following blocks; output transducer, buffer, time delay, switch, oscillator.
b) Calculate the frequency with which the oscillator will oscillate. Give your answer in Hz.
c) State what point C does when point B is
  i) high,
  ii) low.
d) Estimate how long A remains high after the switch is released. (100 s, 10 s, 1 s or 0.1 s?)
e) When the switch has been released for a long time, which of the bulbs is lit?
f) Roughly how many times will bulb U come on after the switch is released?

4 A circuit contains a push switch as its input and a speaker as its output. Whenever the switch is pressed, the speaker must make a sound immediately. When the switch is released, the sound must continue for about 20 s. The sound has a frequency of about 500 Hz.
a) Draw a block diagram for the circuit, using the following blocks; speaker, buffer, oscillator, switch, time delay.
b) Draw circuit diagrams for the following blocks, showing component values for resistors and capacitors.
  i) the oscillator,
  ii) the switch and time delay.
c) Draw a circuit diagram of the whole system.

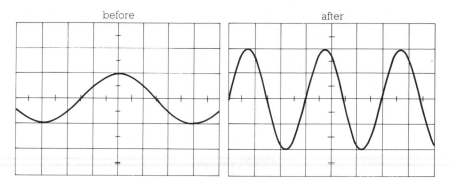

Figure 9.17 Question 5   **Figure 9.18** Question 5

5 A pupil is studying the waveform from a circuit with the help of a CRO. When the timebase is set at 5 ms/cm and the vertical amplifier is set at 2 V/cm, the trace is as shown in figure 9.17.
   a) What is the amplitude of the waveform?
   b) What is the period of the waveform?
   c) While her back is turned, her teacher changes the timebase and vertical amplifier settings so that the trace appears as shown in figure 9.18. What did the teacher change the settings to?

10
Pulses

The output of an oscillator changes state regularly. It starts off low. Then it goes high. Then it goes low again. Each time the output of the oscillator goes high and returns low it emits a **pulse**. Figure 10.1 shows a train of six pulses.

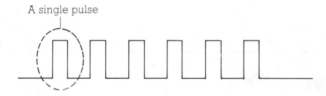

A single pulse

Figure 10.1 A train of pulses

Monostables

It is sometimes useful to be able to create single pulses. Systems which do this are called **monostables** or **triggered pulse generators**.

A circuit symbol for a monostable is shown in figure 10.2, together with a **timing diagram** for it. The output (Q) is normally 0. Each time that the input (T) falls from 1 to 0 Q goes to 1 for RC ms. The system is **triggered** by the **falling edge** fed into T. Provided that T goes back up to 1 before Q has finished producing its single pulse, the pulse length (RC) will be independent of what T does. The triangle at the T terminal tells you that it is **edge-triggered**; the circle next to the triangle tells you that falling edges (i.e. 1 to 0) do the triggering.

Figure 10.2 Timing diagram for a monostable

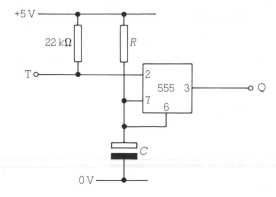

Figure 10.3 A 555 monostable

The 555 monostable

Figure 10.3 shows how a 555 IC has to be connected to make a monostable. The trigger input (T) is pin 2 and the pulse output (Q) is pin 3. The timing components R and C are connected between the two supply rails. Each time that the IC is triggered it discharges the capacitor and lets it charge up through the resistor. When pin 6 gets to the upper trip point of the 555's Schmitt trigger NOT gate it makes Q go to 0 again.

The pulse length will be *RC* ms
if *R* is in kΩ and *C* is in μF.

A 555 monostable in action is shown in figure 10.4. An LDR is used to trigger the system. Normally the LDR is in the light, so that it has a small resistance, pulling T high. If the LDR is momentarily covered over T will get pushed low and the monostable will be triggered.

Q will go high straight away, making the LED glow. After a time interval of $100 \times 47 = 4700$ ms or 4.7 s Q will go low again and the LED will go off.

One of the oddities of the 555 monostable is that if T is kept low then Q will stay high. So it is important that it is triggered by a relatively short pulse.

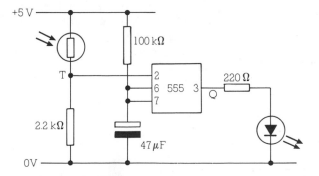

Figure 10.4 A light-level triggered pulse generator

Pulsing waveforms

A monostable can be used to force an oscillator to produce a train of pulses for a fixed length of time. It pulses the waveform of the oscillator on and then off again.

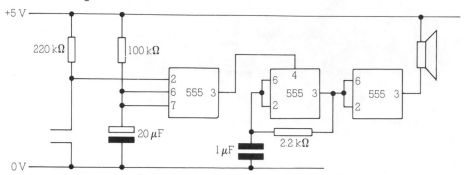

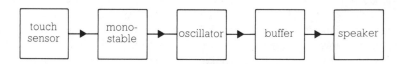

Figure 10.5 A doorbell circuit

For example, the circuit of figure 10.5 could act as a doorbell. Each time that the contacts were briefly touched by a bare finger (which has a resistance of about 50 kΩ) the monostable output would go high for 2 s. This would switch the 200 Hz oscillator on, so that a 2 s burst of sound was produced by the speaker.

Figure 10.6 shows a block diagram of the system. Note how each block has a different function although **three** can be assembled from a 555 IC.

```
touch   →  mono-   →  oscillator  →  buffer  →  speaker
sensor     stable
```

Figure 10.6 Block diagram for figure 10.5

Electronic systems are better

The electronic doorbell system has a number of advantages over its mechanical equivalent.

It is probably cheaper to build, consuming fewer raw materials. It will certainly be much smaller. The bulkiest parts will be the speaker and the battery, but it should be possible to build the whole system into a door with ease. The circuit uses a touch switch, with no moving parts to wear out, so the system should have a long and trouble-free lifetime. Indeed, if CMOS 555 ICs are used the battery will only have to provide current while the speaker is making a noise. So it will have a long lifetime as well. Finally, the tone of the sound could be easily adjusted to suit the muscial tastes of the householder. (Have you noticed that all doorbells appear to sound the same?)

Oscillating oscillators

The output of one oscillator can be used to control the operation of another oscillator. This can produce some very interesting sounds if the resulting waveform is fed into a speaker.

Consider the block diagram of figure 10.7. A slow oscillator switches a fast oscillator on and off. The output of the fast oscillator is fed into a speaker via a buffer. So the speaker produces regular bursts of sound; it bleeps continuously.

Figure 10.7 Block diagram for figure 10.8

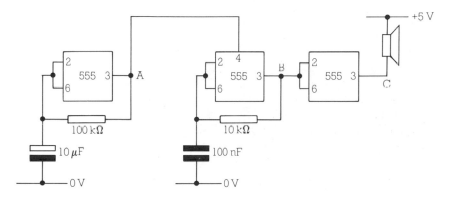

Figure 10.8 A pulsed oscillator

The circuit of figure 10.8 has a 0.5 Hz oscillator on the left. Its output (A) alternates between 1 and 0, spending 1 s in each state. While A is high the fast oscillator will feed a 500 Hz waveform out of B, but while A is low B will be held low. So C will spend half its time feeding a train of 500 pulses into the speaker and the rest of its time staying high.

QUESTIONS

1   A burglar alarm circuit contains an LDR and a buzzer. The circuit is placed in a dark room at night. If a burglar shines a torch on the LDR for an instant, the buzzer must make a sound for a minute.
   a)   Draw a block diagram for the circuit.
   b)   Use the block diagram to help you explain how the circuit works.
   c)   Draw a circuit diagram for the system, showing suitable component values.

2   An electronic egg-timer will make an LED glow immediately a finger is placed on a pair of contacts. Three minutes later the LED goes off and a pulsed tone is produced by the system until the contacts are touched again. State, and explain, in what ways the egg-timer is likely to be superior to its mechanical equivalent.

3 This question is about the circuit shown in figure 10.9. It contains a
monostable.
a) What is the state of T when the switch is
   i) open,
   ii) closed?
b) What happens to A when T is
   i) 1 all of the time,
   ii) 0 all of the time,
   iii) falls from 1 to 0 briefly?
c) When the switch is pressed and immediately released the bulb
comes on for a short space of time. For how long is the bulb lit?
d) What does the bulb do if the switch is pressed and held closed?

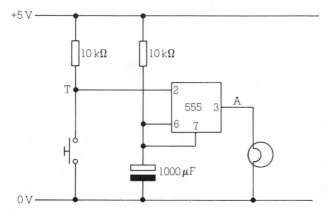

Figure 10.9 Question 1

Revision questions for Section B

1 This question is about the circuit of figure B.1.
- a) What type of capacitor is being used?
- b) What is the voltage at A when the switch is pressed?
- c) Sketch a graph to show how the voltage at A changes with time when the switch is pressed and released.
- d) The LED stays on for 7 s after the switch is released. For how long would it stay on if
    - i) a 1000 μF capacitor had been used,
    - ii) a 10 kΩ resistor had been used?

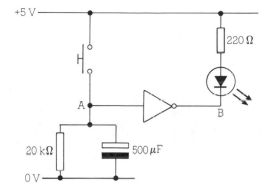

Figure B.1 Question 1

2 An oscilloscope is used to study the waveform of an oscillator. The trace on the screen is as shown in figure B.2 when the vertical amplifier is set at 2 V/cm and the timebase is set at 10 ms/cm.
- a) For how long does the waveform stay high in one cycle?
- b) What is the value of the period of the waveform?
- c) What is the frequency of the waveform?
- d) Show what the trace on the screen would be like if the timebase was set at 5 ms/cm and the vertical amplifier was set at 5 V/cm.

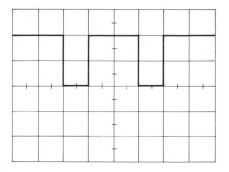

Figure B.2 Question 2

3 The 555 IC shown in figure B.3 oscillates with a frequency of 1 Hz.
a) Describe, as precisely as you can, what the bulb does.
b) The 500 μF capacitor is replaced with a 5 μF one.
   i) What is the frequency of the oscillator now?
   ii) What does the bulb appear to do?
c) Draw a circuit diagram to show how you would adapt the circuit of figure B.3 so that it emitted a sound whose pitch depended on the amount of light that hit an LDR.

4 Study the block diagram shown in figure B.4. The system is supposed to emit bursts of sound at regular intervals.
a) Suggest a suitable output transducer for the system.
b) Calculate the periods of the waveforms at A and B.
c) Sketch waveforms for the signals at points A and B in the system. Mark both the voltage and time axes carefully.
d) How does the state of C depend on the state of B when A is
   i) 1,
   ii) 0?
e) How long is each burst of sound? What is its frequency?

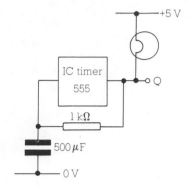

Figure B.3 Question 3

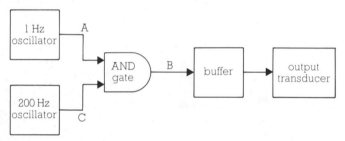

Figure B.4 Question 4

5 The monostable of figure B.5 feeds out a 10 s pulse when it is triggered.
a) Explain what you have to do to T to trigger the monostable.
b) What does Q do when the system is triggered?
c) How would you alter the circuit so that the pulse fed out was 20 s long?

6 The graph in figure B.6 shows the transfer characteristic of a system. V_{IN} is the input voltage and V_{OUT} is the output voltage.
a) Name a device which has this sort of characteristic. Draw its circuit symbol.
b) What is the value of its
   i) upper trip point,
   ii) lower trip point?
c) What is the value of V_{OUT} when the output is
   i) low,
   ii) high?
d) Draw a circuit diagram to show how the device could be made into an oscillator with the help of a resistor and a capacitor.

7 State, with reasons, why a fire alarm system for a house which used an electronic system would be better than one which used another method.

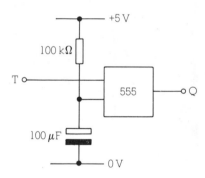

Figure B.5 Question 5

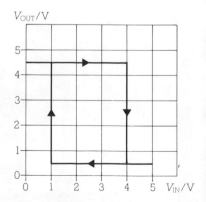

Figure B.6 Question 6

Storing information

This RAM integrated circuit can remember 288 000 bits of information. It is only a few millimetres across. *(IBM)*

11
Bistables

Electronic systems can be used to process **information**. They can take in signals from the outside world via input transducers such as LDRs, switches and contacts. **Those signals contain information.** The signals can then be processed by a logic system (a collection of NOT, AND and OR gates). The output from the logic system can then be used to drive an output transducer (such as an LED, a motor or a speaker) via a suitable buffer.

In all of the systems you have met so far, the state of the output depends on the information going into the inputs, perhaps with a time delay. You are now going to meet systems which can remember the information which was fed into them some time in the past. They are called **bistables**, and they are the box of tricks at the heart of large electronic memories.

The NOT gate bistable

Figure 11.1 contains two NOT gates. The output of each gate is fed into the input of the other gate. (This form of connection is sometimes called **cross-coupling** for obvious reasons.) A 555 IC buffers the output Q so that it can be monitored with an LED.

The system is able to remember which of the two switches was last pressed.

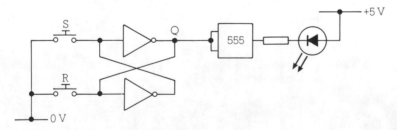

Figure 11.1 Cross-coupled NOT gates

Set and reset

Suppose that S is pressed. This forces Q high, making the LED glow. When S is released, Q stays high so the LED continues to glow. If R is now pressed, Q will be forced low and the LED will go off. It will stay off when R is released.

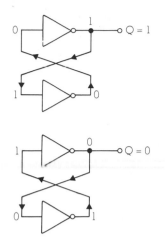

Figure 11.2 The two stable states of a NOT gate bistable

When Q is 1 the bistable is **set**. When Q is 0 the bistable is **reset**. The S and R switches can be used to set or reset the bistable. Information can be fed into the bistable via those switches and remembered. The system can remember, or store, one **bit** of information; the output Q can be 1 or 0.

For example, a NOT gate bistable could be used to store some information about the presence or absence of a pupil at school. At registration the bistable could be set if she was present, otherwise it would be reset. So Q = 1 means that Anne is in school and Q = 0 means that she is absent.

Two stable states

How does the NOT gate bistable work? It has two stable states, hence the name 'bistable.' The output will sit quite happily with Q = 1. It is also quite happy to sit with Q = 0. Both output states are stable.

If you consult figure 11.2, you will see why. The upper circuit has Q = 1. So the output of the bottom NOT gate must be 0. So the output of the top gate must be 1. So Q stays at 1.

The lower circuit has Q = 0. The output of the bottom gate must therefore be 1. So the output of the top gate must be 0. Q stays quite happily at 0.

The two switches in figure 11.1 are used to force Q into either of its stable states. They do it by pulling the output of one or other of the NOT gates low by connecting it directly to the 0 V supply rail. This is brutal but effective.

The NOR gate bistable

NOT gate bistables are not much used because they can only be set or reset with switches. It often happens that you want a bistable to be set or reset by another electronic system rather than by a human being. In that case you need a bistable which has inputs as well as outputs. The **NOR gate bistable** is such a system.

Timing diagrams

A NOR gate bistable is shown in figure 11.3. There is also a **timing diagram** which describes what it does. The timing diagram is three voltage-time graphs placed one on top of the other. It shows what happens to Q as S and R are changed with time. S and R are the set and reset inputs of the bistable. Q is the output.

The instant that S goes high, Q goes high. It stays high until the instant that R goes high, whereupon it goes low and stays low. This is shown in the timing diagram and summarised in the truth table.

Q is stable when S and R are 0.
If S is 1 then Q is set to 1.
If R is 1 then Q is reset to 0.

(We are going to ignore what happens when both S and R are 1 because it is a condition which is always avoidable in practice.)

How it works

The four diagrams of figure 11.4 illustrate how a NOR gate bistable works. It will help if you bear in mind that the output of a NOR gate is 0 if any of its inputs are 1.

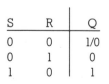

S	R	Q
0	0	1/0
0	1	0
1	0	1

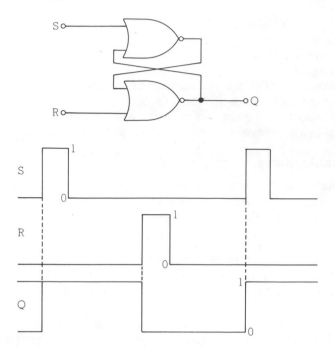

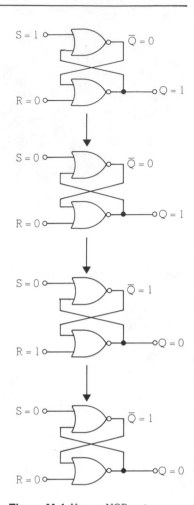

Figure 11.3 Timing diagram for a NOR gate bistable

Start at the top with S = 1 and R = 0. $\overline{Q}$ (pronounced 'q-bar') must therefore be 0. So Q must be 1 and the bistable is set.

In the second diagram S has been lowered to 0. This has no effect on Q or $\overline{Q}$, so Q remains a 1.

R is raised to 1 in the third diagram. This immediately makes Q a 0, allowing $\overline{Q}$ to be a 1. So the bistable has been reset.

Finally R is lowered back to 0. This has no effect on Q or $\overline{Q}$, so the system remains reset.

Figure 11.4 How a NOR gate bistable goes from one state to another

On-off touch switch

The system shown in figure 11.5 uses a NOR gate bistable to remember which of a pair of contacts was last touched by a finger. If the ON contacts are briefly touched the bistable is set, turning the mains light bulb on via a relay. It will stay on until the OFF contacts are briefly touched. Note how a pair of 100 kΩ pull-down resistors keep both S and R low when neither sets of contacts are being touched.

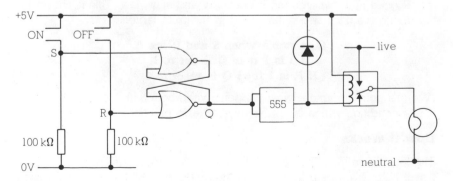

Figure 11.5 On-off touch switch

QUESTIONS

1 Copy and complete the following statements. They refer to the bistable of figure 11.3.

When the bistable is set, Q is .... To set the bistable, S = ... and R = ... When the bistable is reset, Q is .... To reset the bistable S = ... and R = ... The bistable remembers which of S or R was last held at ....

2 This question is about the circuit shown in figure 11.6. It is a simple burglar alarm system. The two switches marked W and D are built into the frames of a window and a door respectively. Both switches close when the door and window are closed.

a) Suppose that both the door and window are closed. What is the state of S?

b) If the SPDT switch is now used to hold R high for a moment, what is the state of Q?

c) What does the buzzer do now?

d) If R is held at 0 by the SPDT switch and the door is opened, what happens to the buzzer?

e) What happens to the buzzer if the door is closed again?

f) Draw a circuit diagram to show how the system could be expanded to monitor two windows and two doors.

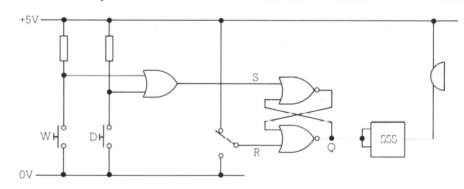

Figure 11.6 Question 2

3 NAND gates can be used to make bistables, as shown in figure 11.7. This question will help you to analyse their behaviour.

a) Write down the truth table for a NAND gate.

b) Complete the truth table shown below.

$\bar{S}$	$\bar{R}$	Q	$\bar{Q}$
0	1	?	?
1	0	?	?

c) When both $\bar{S}$ and $\bar{R}$ are both 1, can Q be
  i) a 0?
  ii) a 1?

d) What do you have to do to
  i) set the bistable,
  ii) reset the bistable?

e) Copy and complete the timing diagram of figure 11.7.

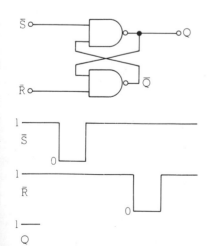

Figure 11.7 Question 3

12
Remembering words

The single bistable of figure 12.1 can store one **bit** of information. That bit can be a 1 or a 0 and it is fed out at Q. A useful system has to be able to store more than one bit. It has to be able to remember a **word**.

Binary words

Suppose that we want an electronic system to store some information about four pupils called David, Charles, Beatrice and Anne. We want to record if they are present or absent on a particular day. Four bistables will be needed, one for each pupil. If the bistable outputs are labelled D, C, B and A, then the **four bit binary word** DCBA will hold the information.

Storing words

A circuit which could be used to obtain and store this word is shown in figure 12.2. A single switch is pressed to reset all four bistables at the start of the school day. As each pupil arrives they press their switch to set a bistable. So if DCBA = 1110, then David, Charles and Beatrice are present but Anne is absent.

Coding information

A particular system can be used to store many types of information. For example, instead of recording the attendance of four different pupils on one particular day, the circuit of figure 12.2 could be used to record the attendance of a single pupil on four different days. So if DCBA = 1110 it could mean that Edward was present on Monday, Tuesday and Wednesday but absent on Thursday.

If you want to keep track of all five pupils for one week (Monday to Friday) you need to store a twenty five bit word. The word will be a string of 1's and 0's, twenty-five in all, and will depend on how the information has been **coded** into it. So the first five bits of the word could represent the attendance of all five pupils on Monday. Or they could represent the attendance record of just one pupil for the whole week.

In order to extract the information stored in a binary word you need to know how it is coded.

The system of figure 12.2 has one great drawback. Each bit can be set at any time. So it cannot distinguish between a pupil who arrived at school on time and one who was late. Both will press their switches when they

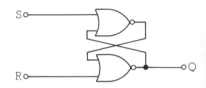

Figure 12.1 A one bit memory

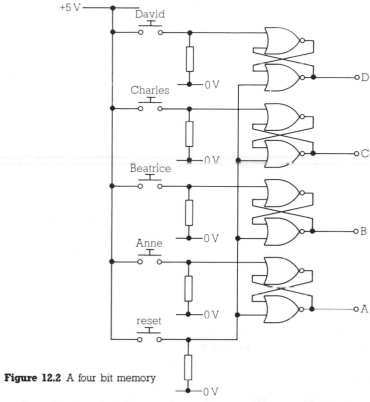

Figure 12.2 A four bit memory

arrive. An improved system would only allow its stored word to be changed at a particular time. This requires the use of a **clocked bistable** or **flip-flop**.

The D latch

A simple clocked bistable known as a **D latch** is shown in figure 12.3. It has one output, Q. So it can store one bit. The D input is used to feed in the bit to be stored. The ST (short for STROBE) input is used to tell the latch when to make Q the same as D.

When ST is 1 then Q is the same as D.
When ST is 0 then Q is frozen.

For example, suppose that you want to make Q a 1. First you hold D at 1. This will have no effect on Q if ST is 0. Then you pull ST up to 1. This will allow Q to become the same as D. Finally, you let ST go low again so that Q cannot change.

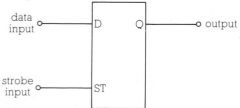

Figure 12.3 A D latch

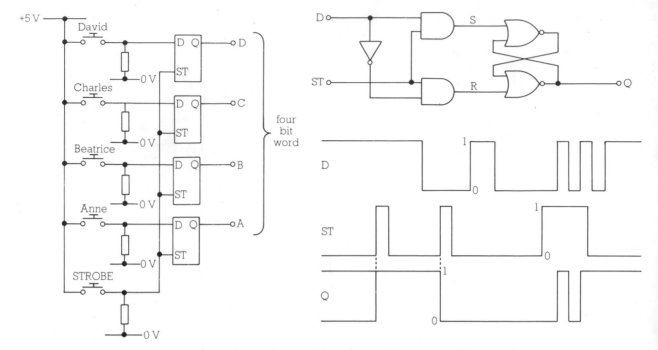

Figure 12.4 Using D latches to store a word

Figure 12.5 A D latch made from logic gates

Timed data capture

An improved registration system which uses D latches instead of NOR gate bistables is shown in figure 12.4. Imagine that it is being used to register the presence or absence of the four pupils David, Charles, Beatrice and Anne. Each pupil that is present presses his switch and keeps it pressed. At 9 o'clock exactly the teacher briefly presses the STROBE switch. The four outputs D, C, B and A now hold information about the presence of the four pupils at 9 o'clock. So if DCBA = 0111 David was absent (or late) but the other three were there.

Inside the D latch

Figure 12.5 shows how a NOR gate bistable can be converted into a D latch with the help of two AND gates and a NOT gate. The timing diagram shows how Q can be set and reset with the D and ST inputs. Notice how Q can only change when ST is 1, and that when it does change it is always the same as D.

The truth table for the three gates in front of the bistable is shown below.

ST	D	S	R
0	0	0	0
0	1	0	0
1	0	0	1
1	1	1	0

While ST is 0 both S and R are 0. So the output of the bistable is stable. To set the output S has to be 1, requiring ST and D to be 1. Similarly, to reset the bistable R has to be 1, requiring ST to be 1 and D to be 0.

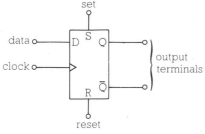

Figure 12.6 A D flip-flop

The D flip-flop

Although a D latch is relatively complex inside, it is quite easy to use. You place the bit that you want to remember at the D terminal. Then you pulse ST high for an instant. The bit will then be held at Q until the next time that ST is pulsed.

The **D flip-flop** will do the same job as a D latch, as well as several others (such as making binary counters). The only difference between the D latch and the D flip-flop is that the latter is **edge-triggered**. So you use a **rising edge** (a change from 0 to 1) rather than a pulse (0 to 1 to 0) to instruct it to store a new bit.

The circuit symbol of a D flip-flop is shown in figure 12.6. Although it looks complicated (it has four inputs and two outputs) it is relatively easy to use. You get two of these flip-flops on a single 4013 IC as shown in figure 12.7.

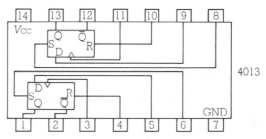

Figure 12.7 The 4013 IC

S and R inputs

The S and R inputs are used to set and reset the flip-flop. They are **active-high inputs** so they are normally held at 0 by pull-down resistors. To set Q to 1 or reset it to 0 the S or R inputs have to be pulled up to 1. This is shown in the truth table below.

S	R	Q
0	0	1/0
0	1	0
1	0	1

(It is not a good idea to try and set and reset the flip-flop at the same time, so we have ignored the S = 1, R = 1 state.)

So a D flip-flop can be made to behave just like a NOR gate bistable if it is connected as shown in figure 12.8. Note how the unused inputs (DATA and CLOCK) have been made 0; it is important that unused inputs to a CMOS IC are held at either 1 or 0.

Edge triggering

A D flip-flop captures data at the instant that its CLOCK terminal rises from 0 to 1.

Suppose that you want to store a 1 in the flip-flop. You first hold the DATA or D terminal at 1. Then you raise the CLOCK or CK terminal from 0 to 1. This makes Q a 1 and $\overline{Q}$ a 0.

To store a 0 in the flip-flop you follow the same procedure but holding D at 0. As soon as CK rises from 0 to 1 Q becomes 0 and $\overline{Q}$ becomes 1.

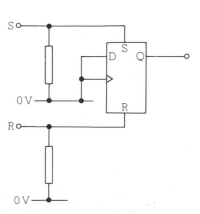

Figure 12.8 Using the set and reset terminals of a D flip-flop

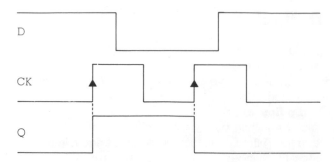

Figure 12.9 Timing diagram for a D flip-flop

Q always has the opposite state to Q̄.
Q becomes the same as D when CK rises from 0 to 1.

Of course, both S and R must be held at 0 when a flip-flop is used this way. If S is 1, then Q will be 1 regardless of what you do to the D and CK terminals!

The timing diagram of figure 12.9 summarises the behaviour of a D flip-flop when it is **clocked**. Notice that Q is frozen except during the instant that CK rises from 0 to 1.

Clean edges

Because it is triggered by a change of voltage a D flip-flop is very fussy about the shape of its clock pulses. They must change very quickly.

Clock pulses which come from electronic systems are usually very good in this respect. Pulses obtained from switches are not so good. They need to be **cleaned up** before being fed into the CK input of a D flip-flop.

Switch bounce

Figure 12.10 shows the type of circuit which might be used to obtain a rising edge with the help of a switch.

V_{OUT} will rise rapidly from 0 V to +5 V each time the switch is pressed. The switch contacts will **bounce** when the switch is pressed or released. This only lasts for a few milliseconds, but it can generate several pulses where you only intended to generate one. So when you close the switch V_{OUT} will oscillate between 0 V and +5 V a few times before settling down to +5 V. A similar thing happens when the switch is released. Several falling edges are produced instead of just one.

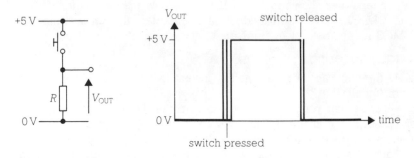

Figure 12.10 Pulses generated by switch bounce

Switch de-bouncing

A capacitor can be used to eliminate the unwanted spikes generated when a switch is opened or closed. Look at figure 12.11. When the switch is closed, the first connection of its contacts allows the capacitor to charge up instantly. The capacitor is then able to keep V_{OUT} at about +5 V during subsequent bounces of the contacts.

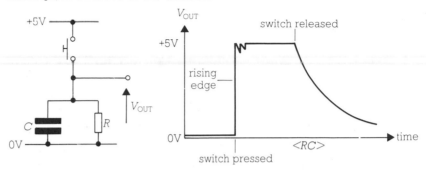

Figure 12.11 Using a capacitor to remove unwanted pulses

So a single rising edge is produced when the switch is closed. Unfortunately, V_{OUT} falls slowly from +5 V to 0 V when the switch is released. The falling edge is not very crisp.

Figure 12.12 shows how a Schmitt trigger NOT gate can be used to obtain clean rising and falling edges. The capacitor masks the effect of any bouncing of the contacts, so V_{OUT} rises from 0 V to +5 V and stays there as soon as the switch is pressed. When the switch is released there is a delay of about RC ms before V_{OUT} drops cleanly back to 0 V.

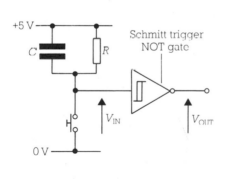

Figure 12.12 A switch de-bounce circuit

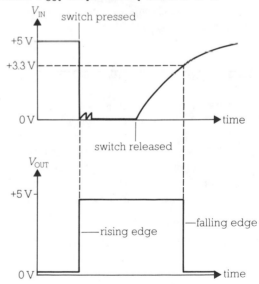

Heads and tails

Suppose that you feed a square wave into the D input of a flip-flop. D alternates between 1 and 0, spending equal amounts of time in each state. So every time that a rising edge is fed into the CK terminal there is an equal chance that Q will be a 1 or a 0. If a 1 represents heads and a 0 represents tails the system behaves like an electronic coin tosser.

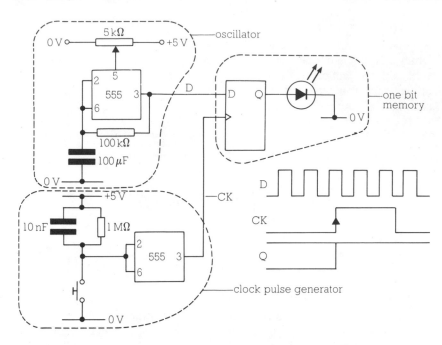

Figure 12.13 A heads-and-tails generator

A circuit which behaves this way is shown in figure 12.13. The oscillator produces a 50 Hz square wave. The **clock pulse generator** produces a clean rising edge every time that the switch is pressed. An LED shows if D was 1 or 0 when the switch was last pressed. The D flip-flop is acting as a **one bit memory**.

Remembering numbers

A single flip-flop can be used to store one bit. To store two bits you would need two flip-flops, one for each bit. Figure 12.14 shows how D flip-flops can be arranged to store a four bit word.

The system is known as a **four bit latch**, even though it uses D flip-flops instead of D latches! The word to be stored is placed on the four D inputs of the flip-flops. When the common clock input is raised from 0 to 1 that word is stored at the four Q outputs of the flip-flops.

Binary coded decimal

The word that the system remembers could represent many different things. In particular, it could represent any number between 0 and 9. The standard method of representing such numbers with a four bit binary word is called **binary coded decimal**, usually shortened to **BCD**.

This is how BCD works. Each bit of the word is given a different label. The labels are D, C, B and A. So the whole word is DCBA. The number which this represents is given by this formula.

$$\textbf{number} = \textbf{8} \times \textbf{D} + \textbf{4} \times \textbf{C} + \textbf{2} \times \textbf{B} + \textbf{1} \times \textbf{A}$$

For example, suppose that DCBA = 0101. This represents the number $8 \times 0 + 4 \times 1 + 2 \times 0 + 1 \times 1 = 0 + 4 + 0 + 1 = 5$.

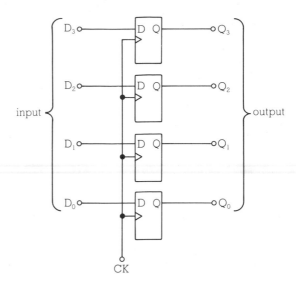

Figure 12.14 A four bit latch

BCD is used for storing numbers in electronic systems. For example the number 741 could be stored with three four bit latches, one for each of the numbers 7, 4 and 1. The whole word stored would be 0111 0100 0001.

Displaying numbers

As well as storing numbers, electronic systems often have to be able to display them. The most user-friendly type of display uses a **seven segment LED**. This uses seven LEDs in the form of bars arranged to make a figure of eight, as shown in figure 12.15. Each LED can be separately lit, so that all of the numbers between 0 and 9 can be generated.

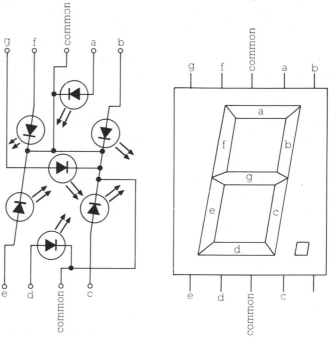

Figure 12.15 A seven segment LED

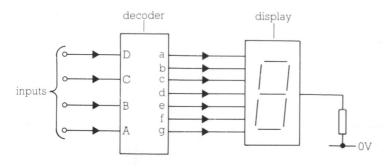

Figure 12.16 Using a decoder

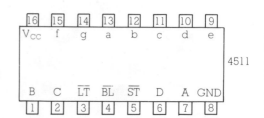

Figure 12.17 The 4511 decoder IC

A current-limiting resistor will normally be connected to each of the LED anodes. A single resistor connected to the joined cathodes can be used instead (see figure 12.16). The use of seven separate resistors means that all of the numbers displayed will have the same brightness.

Decoders

A number between 0 and 9 is represented by a four bit word. But a seven segment LED needs seven bits of information to display a number, one bit for each segment. So we need a **BCD to seven segment decoder** between the LED display and the rest of the system. As shown in figure 12.16 it takes in a four bit word **DCBA** and feeds out the appropriate seven bit word **gfedcba**. The decoder is obviously a complex logic system!

The 4511 IC

The pinout of the 4511 decoder IC is shown in figure 12.17. It has buffered outputs, so it can push plenty of current through the LEDs attached to its outputs.

The IC contains four D latches which can store the four bit word DCBA being fed into it. If the $\overline{ST}$ pin is held low then the output word (gfedcba) will change as the input word (DCBA) is changed. When $\overline{ST}$ is high, the seven outputs are frozen. **So the IC can store a number as well as decode it for the display.**

The IC has two special control pins. They are active-low inputs so they are normally held at 1. If $\overline{BL}$ goes low then all seven segments of the display are blanked off. (They go out.) On the other hand, if $\overline{LT}$ goes low all of the segments are lit. This is useful for testing the system.

A two bit latched decoder

The circuit diagram of a decoder which can only handle numbers between 0 and 3 is shown in figure 12.18. The two flip-flops on the left store a two bit word BA when CK rises from 0 to 1. The logic gates generate the signals needed to get the right patterns on an LED display according the the truth table below.

number	B	A	a	b	c	d	e	f	g
0	0	0	1	1	1	1	1	1	0
1	0	1	0	1	1	0	0	0	0
2	1	0	1	1	0	1	1	0	1
3	1	1	1	1	1	1	0	0	1

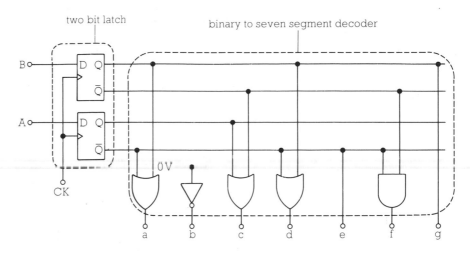

Figure 12.18 A two bit binary to seven segment decoder and latch

QUESTIONS

1 a)   Draw the circuit symbol of a D flip-flop. Label the CK, D, Q, $\overline{Q}$, R and S terminals.

b)   Copy and complete the following statements.
The .. and .. terminals are used to set or reset the flip-flop. When only S is 1, Q is ... and $\overline{Q}$ is .... When only R is 1, Q is ... and $\overline{Q}$ is .... In normal use both S and R are held at .... When CK goes from ... to ... the bit at ... is stored at .... The flip-flop is triggered by ...... edges.

c)   Copy and complete the timing diagram shown in figure 12.19.

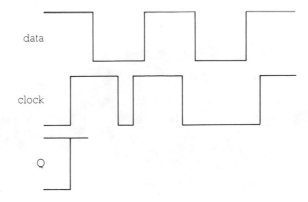

Figure 12.19 Question 1

2 Draw the circuit diagram of a system which could de-bounce the pulses generated by a switch. Explain how it works.

3 Each of these four bit words represent a number coded in BCD. What are the numbers?

1000	0100	0010	0001	0011
1001	0111	0000	0110	0101

4 Write down the four bit words representing the numbers from 0 to 9 in BCD.

5 Draw a seven segment LED. Label each segment with a letter from a to g. Work out which segments have to be lit to display each of the numbers between 0 and 9.

6 What is a decoder? Draw a diagram to show how one could be used to let an LED display show the number stored by a four bit latch.

7 Draw a circuit diagram to show how D flip-flops could be used to store a three bit word XYZ. Explain how to store the word in the circuit.

8 Draw a circuit diagram to show how a pair of 4511 ICs and seven segment LEDs could be used to store and display numbers between 0 and 99. Explain how to store the number in your circuit.

A word processing system. It contains enough RAM to hold over half a million bits of information. *(Amstrad)*

13
Semiconductor memories

A **semiconductor memory** is a solid-state electronic system which can store binary words. For example, the 6116 IC can store 2048 eight bit words at a cost of about ¼ p per word. The actual integrated circuit (rather than the package which contains it) is about a millimetre across, so the 6116 can store a lot of information in a very small space. This ability to cram enormously complex circuits into tiny areas of silicon means that semiconductor memories are going to become very important in the future.

Organization of memory

An electronic memory is a bit like a card filing system. You can think of a 2048 word memory as a pile of 2048 blank cards. Each card represents a **memory location**. It has eight spaces on it, one for each **bit** of the **word** that it can store.

A pencil can be used to **write** a word onto the card. You can subsequently **read** the word off the card. An eraser can be used to rub out the word.

Address

If you were able spread all of the cards out on the floor you could look at all 2048 words at once. If you want to read one particular word, you need some means of finding the card (**location**) that it is written on. Each card must therefore have an **address** written on it. This could be a number between 0 and 2047, with each card having a different address. Furthermore, it would be sensible to stack the cards in a pile with the addresses in order. So if the address of the bottom card was 0 and that of the top card was 2047, finding any card in the stack would be straightforward. Figure 13.1 shows a representation of such a memory system.

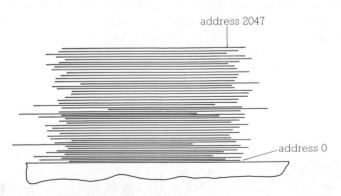

Figure 13.1 A pile of cards

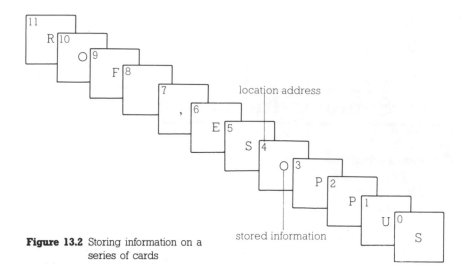

Figure 13.2 Storing information on a
series of cards

Finding words

How do you go about using such a memory? Suppose, for example, that
you want to store this sentence in the memory. It contains 73 characters.
(Count them!) Each character can be represented by a seven bit word if
we use the **ASCII** code (the one that everybody uses). We would write
the code for S on the card whose address was 0, the code for u on the
card whose address is 1, the code for p on the card whose address is 2
. . . . ending up with the code for . on the card whose address is 72.
(See figure 13.2.)

The whole sentence would be held in the 73 locations between address
0 and address 72. To read out the sentence from the memory, each
location would have to be read one after the other, starting with the one
whose address is 0.

RAM and ROM

There are two basic types of electronic memory known as **RAM** and
ROM.

RAM stands for **random-access memory**, but it really means **read-and-
write memory**. RAM can have words written into any of its locations, as
well as having words read from them. So the contents of RAM can be
changed by other electronic systems. The words stored in a RAM are lost
when it is disconnected from its power supply. So it cannot be used for
permanent storage of information unless it has a power supply built into it
(and some do).

ROM stands for **read-only-memory**. This means that you cannot write
words into it. You can only read the words stored in its locations. Those
words are usually written into the memory when it is constructed and are
retained in the ROM when its power supply is disconnected. So ROM is
used for the permanent storage of information. Some types of ROM such
as EPROMS can have their contents **erased** wholesale and have fresh
words **programmed** into them.

The difference between RAM and ROM is similar to the difference
between an exercise book and a text book. You have to write words into

an exercise book before they can be read, whereas the text book arrives with the words permanently printed in it.

Using ROM

A ROM will usually have a large number of terminals, but they will fall into three categories. Figure 13.3 shows the input and output terminals of a 2048 × 8 bit ROM.

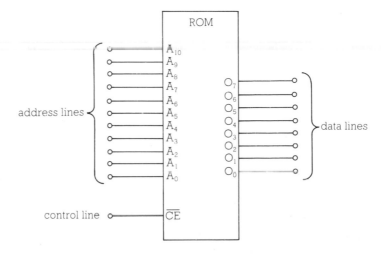

Figure 13.3 A ROM

The **data lines** will feed words out of the ROM. An eight bit ROM will have eight data lines, one for each bit.

The **address lines** are used to select the location whose word is to be fed out of the data lines. The address of a location will be a binary word. A ROM with 11 address lines will have 2^{11} = 2048 different locations.

Finally, there will be some **control lines**. These are used to disable (or disconnect) the ROM's data lines, so that it stops feeding out a word. The $\overline{CE}$ line controls the ROM outputs as shown in the truth table below.

$\overline{CE}$	Data lines
0	Enabled
1	Disabled

So if you want to read a particular word stored in that ROM you have to do the following.

1) Feed the eleven bit word which is the address of the word's location into the address lines.

2) Hold the $\overline{CE}$ (chip enable) line at 0.

3) Read the word off the eight data lines.

Using RAM

Figure 13.4 shows the input and output terminals of a small 4 × 4 bit RAM. It has address and data lines which have the same function as those of a ROM, but it has two control lines instead of just one. One enables the RAM and the other tells it to either feed out a word or read one in.

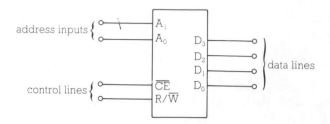

Figure 13.4 A RAM

The function of the control lines is summarised in the truth table below.

$\overline{CE}$	R/$\overline{W}$	Function
0	0	Read in a word and store it
0	1	Feed out a stored word
1	0	Data lines disabled
1	1	Data lines disabled

Writing words

Suppose that you wanted to write the word 0110 into address 10. This the sequence of operations required.

1) Set up the address by making $A_1A_0 = 10$.

2) Make $D_3D_2D_1D_0 = 0110$. This feeds the word that we want to store into the RAM's data lines.

3) Set R/$\overline{W}$ to 0. This instructs the RAM that we want to **write** a word into it.

4) Pulse $\overline{CE}$ to 0 for an instant. When this happens the RAM reads in the four bit word present on the data lines (0110) and stores it at the location whose address is 10.

Reading words

You follow a similar sequence to read words out of the RAM. Suppose that you want to read the word stored in the location whose address is 11.

1) Make $A_1A_0 = 11$.

2) Set R/$\overline{W}$ to 1. This instructs the RAM that we are going to **read** a word from it.

3) Hold $\overline{CE}$ at 0. The RAM feeds out the four bit word that is stored at address 11 on the data lines.

Inside a RAM

Figure 13.5 shows how a 4 × 4 bit RAM can be put together. It consists of three parts. The four **memory stores** hold the words stored in the memory. (Each store is a location.) The **address decoders** select one of the stores according to the two bit address A_1A_0 being fed into the system. Finally, the **control logic** instructs the selected store to either feed out a word or to store one.

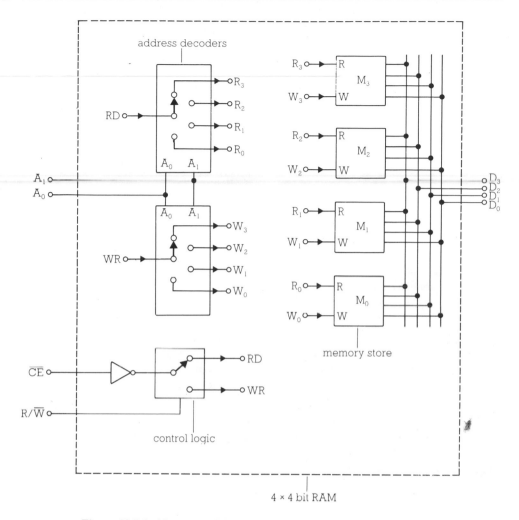

Figure 13.5 Inside a small RAM

The memory store

A single memory store is shown in figure 13.6. Each flip-flop holds one bit of the word being stored. The two control lines (R and W) tell the store what to do according to the truth table shown below.

R	W	Function
0	0	Disable input-output lines
0	1	Write a word into the store
1	0	Read out word in the store

Each time W goes high the four flip-flops store the state of the input-output lines. R controls four **tristates**. Each of these acts like a switch connecting the output of a flip-flop to an input-output line.

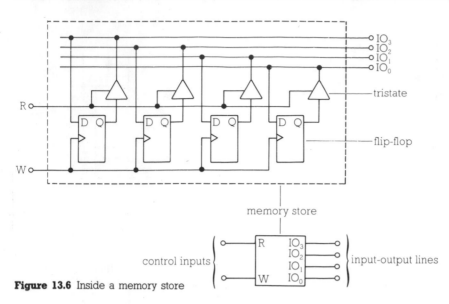

memory store

control inputs { input-output lines

Figure 13.6 Inside a memory store

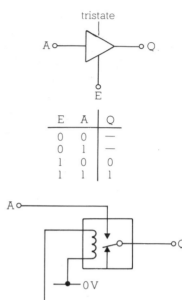

Figure 13.7 A tristate

Tristates

The circuit symbol and truth table of a tristate are shown in figure 13.7. The relay equivalent of a tristate is also shown; it may help you to understand what a tristate does. When E is 0 the tristate output is disabled. This is shown with a − in the truth table.

The memory store needs tristates because it uses the same four lines for both reading data in and feeding it out.

The control logic

The two control signals which enter the RAM ($\overline{CE}$ and R/$\overline{W}$) have to be **decoded** to make the read and write signals (RD and WR) for the memory stores.

$\overline{CE}$	R/$\overline{W}$	RD	WR	Function
0	0	0	1	Write
0	1	1	0	Read
1	0	0	0	Disable
1	1	0	0	Disable

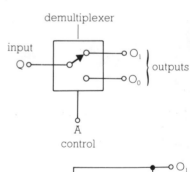

Demultiplexers

The decoding of the control and address lines is done with **demultiplexers**. One of these is shown in figure 13.8. It behaves a bit like a simple telephone exchange, switching the signal at Q to either one of the outputs O_1 or O_0 depending on the state of the control pin A.

A	Q	O_1	O_0
0	0	0	0
0	1	0	1
1	0	0	0
1	1	1	0

The relay version of the demultiplexer shown in figure 13.8 may help.

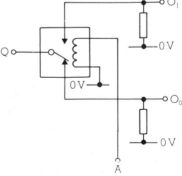

Figure 13.8 A demultiplexer

Clearly, when A is 0 then O_0 will be connected to Q. When A is 1, then O_1 is connected to Q.

The address decoders

The two address decoders of figure 13.5 use demultiplexers to route the read and write signals (RD and WR) from the control logic to one of the memory stores.

Figure 13.9 shows how each of the four output demultiplexers can be made out of three two output ones. The two bit binary word A_1A_0 selects which of the four outputs is made equal to the input O according to the truth table shown below.

Address A_1A_0	O_0	O_1	O_2	O_3
00	Q	0	0	0
01	0	Q	0	0
10	0	0	Q	0
11	0	0	0	Q

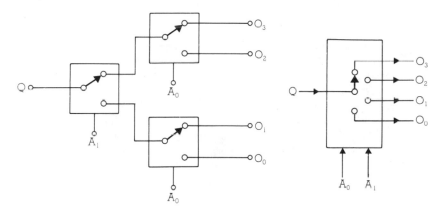

Figure 13.9 A four output demultiplexer

If you add four more two output demultiplexers to the right hand end of figure 13.9 you get an eight output one with three control lines. It should be obvious that it is easy to design large demultiplexers from small ones. Figure 13.10 shows how the simplest demultiplexer can be made from logic gates.

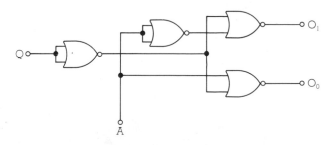

Figure 13.10 A simple demultiplexer from NOR gates

Memories in action

The availability of cheap large electronic memories is going to have an enormous impact on our lives in the near future.

Large RAMs

Although the RAM that we have been studying is a very small system, it should be obvious how to design a larger one. For a start, you need more flip-flops in each memory store, perhaps eight instead of four. (An eight bit word is called a **byte** and is ideal for storing alphanumeric characters.) You might also have 2048 memory stores instead of just four. Although this will require larger demultiplexers to do the address decoding, large demultiplexers are easily made by connecting together a number of smaller ones (see figure 13.9). The design of a large RAM is very straightforward once you have managed to design a small one!

Although the circuit diagram of such a 2048 × 8 bit RAM would be very large, the actual circuit could be etched onto a very small slice of silicon. Provided that the demand for such RAM ICs was high enough, production costs would be low.

Using RAM

What is the point of wanting to make such a large electronic memory? Well, once you have some information stored in electronic form you can **process** it very rapidly. You can get access to any one of the words stored in a 6116 2K RAM in about 0.0002 ms. (**1 K = 1024.**)

For example, the text of this chapter can be stored in a 20 K RAM within a **word processor**. Each character is represented by an eight bit word and stored in one of the 20 × 1024 = 20480 locations of the RAM. So the chapter can be stored in electronic form by ten 6116 ICs.

Speed

Once it is stored this way information can be manipulated very rapidly. Any part of the chapter could be read out of the memory very quickly. Any part could be changed, put somewhere else or deleted. Manipulating the text of the chapter can be done far more rapidly by electronic means than by any other. Suppose that you wanted to replace the word memory with the word brain throughout the whole chapter. It might take you ten minutes to do it with a pen and text book, but a word processor can do it in 15 s! Furthermore, you can manipulate the text in electronic form as many times as you like without degrading its appearance. This is not the case if text is manipulated when stored on paper, unless re-typed!

Fast transfer

Information stored in an electronic memory can be transmitted from one place to another very rapidly. The 20 K of RAM containing this chapter could transfer its contents to another RAM via a telephone line in about ten minutes. This is much faster than you could type it! Telephone lines are relatively slow; they can only transmit about 30 bytes per second. A direct link from one word processor to another could dump the text of this chapter from one to the other in a few seconds.

Information in electronic memories can be processed very rapidly.

Storage density

RAM can only be used for temporary storage of information. Computers and word processors currently use magnetic discs or tape for the permanent storage of information. A small disc might hold 400 K bytes; two such discs can store this whole book. At the moment this method of permanent storage is cheaper and more convenient than using an EPROM, particularly if the information being stored needs altering from time to time.

ROM is a more attractive storage medium for information which is not going to be altered. More and more memory stores can be crammed into a single IC every year, so their storage capacity is improving rapidly. ROMs are not as susceptible to damage as paper or magnetic discs are, and (provided production runs are large enough) their cost can be very low.

Quite soon it may be more cost effective to store long term information in ROMs rather than on paper, particularly if you need access to that information from time to time. If your information is filed on paper, it can take you a long time to find it and if the paper gets filed in the wrong place you may lose it completely. Information that is stored in ROM can be summoned up instantly and cannot be misplaced!

QUESTIONS

1 An electronic memory has three sorts of lines going into it. State what they are and what they are for.

2 Describe the differences between RAM and ROM.

3 This question is about the RAM of figure 13.4.
   a) Describe what you have to do to store the word 1110 in the location whose address is 01.
   b) Describe what you have to do to read the word stored at address 10.

4 The demultiplexer of figure 13.8 is a logic system with two inputs (A and Q) and two outputs (O_1 and O_0).
   a) Draw up its truth table.
   b) Draw a diagram to show how it can be built from two AND gates and a NOT gate.
   c) Show how seven two output demultiplexers can be connected to make an eight output demultiplexer.

5 Figure 13.11 shows an 8 × 3 bit ROM wired up to some switches,

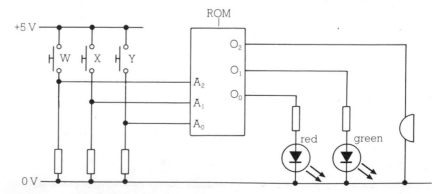

Figure 13.11 Question 5

LEDs and a buzzer. A_2, A_1 and A_0 are the address inputs. O_2, O_1 and O_0 are the outputs.

a) The table shows the words which are stored in the various locations in the ROM.

   i) What do you have to do to the switches to make the buzzer buzz?

   ii) Which LED is lit when only one switch is pressed?

   iii) How many switches have to be pressed to make the red LED glow?

b) Adapt the circuit of figure 13.11 so that it has three switches and three LEDs. You can only store the following words in the ROM; 000, 001, 011 and 111. Draw up a table to show which word must be stored at each location if the number of lit LEDs is to be the same as the number of switches being pressed.

Address $A_2A_1A_0$	Word $O_2O_1O_0$
000	000
001	010
010	010
011	001
100	010
101	001
110	001
111	100

6 State, with reasons, why it would be a good idea for a large hospital to keep the medical records of past patients stored in a series of ROMs rather than on paper filed in the basement.

7 Explain what a tristate does.

8 Modern telephones often contain some electronic memory. Explain what that memory is used for. Must it be RAM or ROM? In what way does the presence of the memory make the telephone more efficient?

9 The circuit of figure 13.12 shows how a 4 × 2 bit ROM has been constructed from a number of components.

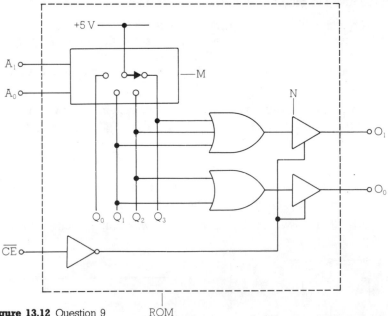

Figure 13.12 Question 9   ROM

a) Name component N. Describe its behaviour with the help of a truth table. What is its function within the ROM?

b) Name component M. What is its function within the ROM?

c) Copy and complete the truth table shown.

d) Work out the two bit words O_1O_0 stored in each of the four locations of the ROM.

A_1	A_0	Q_3	Q_2	Q_1	Q_0
0	0	?	?	?	?
0	1	?	?	?	?
1	0	?	?	?	?
1	1	?	?	?	?

Revision questions for Section C

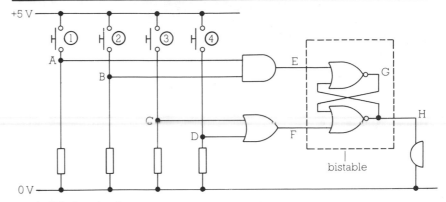

Figure C.1 Question 1

1 The circuit of figure C.1 shows four push switches (labelled 1 to 4) which can be used to control a buzzer.
   a) The part of the circuit in a box is called a bistable. What does the word bistable mean?
   b) State what you have to do to E and F to make H go
       i) high,
       ii) low.
   c) What happens to H if both E and F are low?
   d) What is the state of A when switch 1 is
       i) open,
       ii) closed?
   e) Which switches have to be pressed to make the buzzer work? What happens when they are released?
   f) Which switch can you press to turn off the buzzer? What happens when you release that switch?

2 This question is about the component shown in figure C.2.
   a) What is the component called?
   b) Two of the inputs of the component have been omitted in figure C.2. Redraw the component symbol correctly. State the function of the inputs which were omitted.
   c) If Q is 1, what is $\overline{Q}$?
   d) Copy and complete the timing diagram of figure C.2. Q is initially 0 and $\overline{Q}$ is initially 1.
   e) Explain why the component of figure C.2 is sometimes called a one bit RAM.

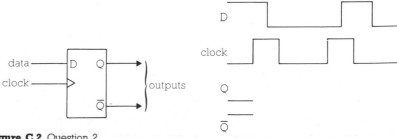

Figure C.2 Question 2

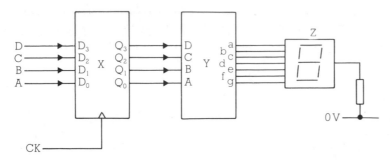

Figure C.3 Question 3

3 The circuit of figure C.3 can store and display a number between 0 and 9.
   a) What is the name of component Z?
   b) By means of drawings, show which segments of Z have to be lit to display the number
       i) 2,   ii) 4,   iii) 6.
   c) What is the name of component Y?
   d) What number will be displayed if the four bit word DCBA is
       i) 0000, ii) 0011, iii) 1001?
   e) What does the word DCBA have to be to display the number
       i)   one,
       ii) seven,
       iii) two?
   f) Component X is a four bit latch. Describe what it does.
   g) Show how you could make component X from four D flip-flops. Label its inputs and outputs.

4 Explain what a ROM is. In what respects is it different from a RAM? Give one example of a suitable use for each of these types of memory.

5 Figure C.4 shows an 8 × 2 bit RAM. The truth table below shows how the function of the memory is controlled by the control lines.

$\overline{CE}$	$R/\overline{W}$	Function
0	0	Read in a word and store it
0	1	Feed out a stored word
1	0	Data lines disabled
1	1	Data lines disabled

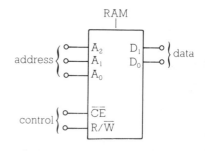

Figure C.4 Question 5

   a) How many locations does the RAM contain?
   b) How long a word can each location hold?
   c) Describe how you would get the RAM to store the word $D_1D_0 = 10$ at the address $A_2A_1A_0 = 110$.
   d) Describe how you would read the word which was stored in the location whose address was $A_2A_1A_0 = 010$.
   e) What does RAM stand for? What happens to the words stored in a RAM when
       i) $\overline{CE}$ is 1,
       ii) the power supply is switched off?

6 Describe and explain the advantages of storing information in electronic memories rather than on paper.

Processing analogue signals

The control desk of a recording studio. The signals from the many microphones are amplified, filtered and mixed before being stored on magnetic tape. *(Pictor International)*

14
Operational amplifiers

This section of the book is going to introduce you to the art of processing **analogue** electronic signals. The systems that you have met so far were designed to process **digital** signals. That is, they only recognise two types of signal, namely those that are high (a 1) or low (a 0). Logic gates, flip-flops, oscillators, monostables and memories are all examples of **digital systems**.

Analogue systems process signals which can have any one of an infinite range of voltages. Figure 14.1 shows voltage-time graphs of typical analogue and digital signals. Notice how the analogue signal has a continuously varying voltage (even going negative) whereas the digital one can only be +5 V or 0 V.

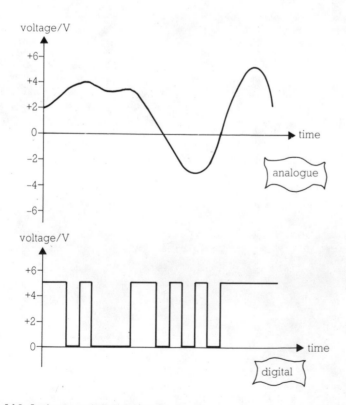

Figure 14.1 Analogue and digital signals

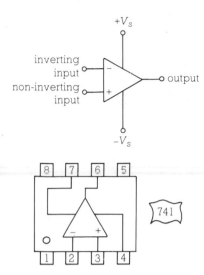

Figure 14.2 An operational
amplifier

Op-amps

Operational amplifiers are the basic building blocks of analogue systems.

The circuit symbol for one is shown in figure 14.2. It looks fairly complicated, with two inputs and one output as well as two supply terminals. The pinout of a **741 IC** is also shown. It contains a single operational amplifier (or **op-amp**). Although many types of op-amp are available, the 741 is the industry standard. As well as being cheap and easy to use, it also has the typical behaviour of an op-amp. So we shall be using it in all of our examples.

Supply rails

Op-amps are designed to run off **split supply rails**. They are labelled $+V_S$ and $-V_S$ in figure 14.2. A 741 IC will operate off supply rails that are between 6 V and 32 V apart. We shall be using $+5$ V and -5 V for $+V_S$ and $-V_S$.

Transfer characteristics

The table below summarises the transfer characteristics of a 741 op-amp which is run off split supply rails of $+5$ V and -5 V.

Input	V_{OUT}
$V_+ < V_-$	-4 V
$V_+ > V_-$	$+4$ V

V_+ is the voltage of the **non-inverting input**. If it is higher than V_- (the voltage of the **inverting input**) then V_{OUT} will go as high as it can. This will be about 1 V lower than $+V_S$ i.e. $+4$ V.

V_{OUT} will go as low as it can (about -4 V) if the inverting input has a higher voltage than that of the non-inverting input.

Figure 14.3 shows the transfer characteristic of an op-amp which has one of its inputs held at a fixed voltage. A voltage divider holds the inverting input at $+2.5$ V and a potentiometer allows the voltage of the non-inverting input to be set at any value between $+5$ V and -5 V.

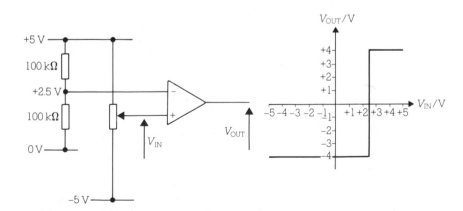

Figure 14.3 Transfer characteristic of an op-amp

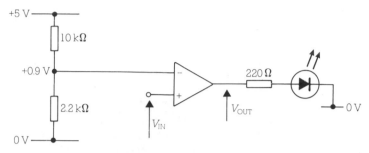

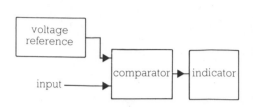

Figure 14.4 A simple voltage indicator

Figure 14.5 Block diagram of a simple voltage indicator

Voltage indicators

Op-amps compare the voltages at their two inputs. They can therefore be used as **voltage indicators**.

The circuit of figure 14.4 acts as a very simple voltmeter. If V_{IN} is above $+0.9$ V the LED will glow. If V_{IN} is below $+0.9$ V the LED will not glow. The whole system gives a very simple (but crude) visual indication of the value of V_{IN}.

A block diagram of the system is shown in figure 14.5. The **voltage reference** is the voltage divider, the **comparator** is the op-amp and the **indicator** is the LED.

How does it work? If V_{IN} is above $+0.9$ V then V_{OUT} will be $+4$ V. So the LED will be forward biased and a current of 9 mA will go through it, making it glow. But if V_{IN} is below $+0.9$ V then V_{OUT} will be -4 V. So the LED will be reverse biased, no current will go through it and it won't glow.

Voltage dividers

Voltage dividers are very useful generators of reference voltages. For example, the 10 kΩ and 2.2 kΩ resistors in figure 14.6 produce a fixed $+0.9$ V from the $+5$ V and 0 V supply rails. This is how they do it.

The total resistance of the voltage divider is $10 + 2.2 = 12.2$ kΩ. The voltage across that resistance is 5 V. So we can use Ohm's Law to calculate how much current goes through it.

$$R = \frac{V}{I} \qquad \begin{array}{l} R = 12.2 \text{ k}\Omega \\ V = 5 \text{ V} \\ I = ? \end{array} \qquad \text{therefore } 12.2 = \frac{5}{I}$$

$$\text{therefore } I = \frac{5}{12.2} = 0.41 \text{ mA}$$

The output of the system is the voltage across the 2.2 kΩ resistor. We can use Ohm's Law to calculate this as we know the values of both the resistance and the current.

$$R = \frac{V}{I} \qquad \begin{array}{l} R = 2.2 \text{ k}\Omega \\ V = ? \\ I = 0.41 \text{ mA} \end{array} \qquad \text{therefore } 2.2 = \frac{V}{0.41}$$

$$\text{therefore } V = 2.2 \times 0.41 = 0.9 \text{ V}$$

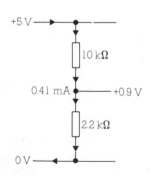

Figure 14.6 Generating a reference voltage with a voltage divider

Notice how we have assumed that the same current goes through both resistors. The input of an op-amp draws a negligible current, so we can make this assumption quite safely.

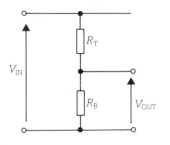

Figure 14.7 Voltage divider symbols

The voltage divider formula

The following rule allows you to calculate the output voltage of a voltage divider. The symbols are defined in figure 14.7.

$$V_{OUT} = \frac{R_B}{R_T + R_B} \times V_{IN}$$

It only works if a negligible current is drawn from the output.

The voltage divider formula can be used to calculate the reference voltage of figure 14.4.

$$V_{OUT} = \frac{2.2 \times 5}{10 + 2.2} = 0.90 \text{ V}$$

Temperature sensing

In order to make an electronic temperature indicator we need a suitable transducer. The circuit of figure 14.8 uses a **thermistor**. This is a resistor whose resistance drops rapidly as its temperature is raised. The graph of figure 14.9 shows the typical behaviour of a thermistor.

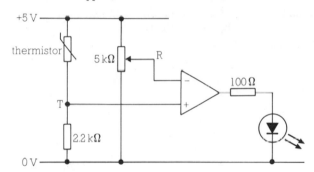

Figure 14.8 An electronic thermometer

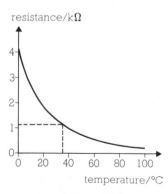

Figure 14.9 How the resistance of a typical thermistor changes with temperature

An electronic thermometer

The electronic thermometer of figure 14.8 has been designed to make the LED come on if the temperature of the thermistor goes above 36 °C. (It could be used to test the temperature of a baby's bath water.) The LED will come on if the voltage at T is higher than the voltage at R. So what voltage do we set up at R with the potentiometer?

The graph of figure 14.9 can be used for the thermistor of figure 14.8. At 36 °C its resistance is 1.1 kΩ. The voltage divider formula can be used to calculate the voltage at T for this resistance.

$$V_{OUT} = \frac{R_B \times V_{IN}}{R_T + R_B} = \frac{2.2 \times 5}{1.1 + 2.2} = 3.3 \text{ V}$$

So we set R to +3.3 V. At low temperatures, when the thermistor has a large resistance, T will be below +3.3 V. So the op-amp's output will be at −4 V and the LED will not glow. But at high temperatures the thermistor's low resistance will pull T above +3.3 V, the op-amp output will be at +4 V and the LED will glow.

Bargraph voltmeters

Figure 14.10 shows how a number of op-amps can be used to make a voltmeter which has a visual output. The output is a row of four LEDs, and the number of LEDs which are glowing indicates the value of V_{IN}.

Each LED is controlled by an op-amp. That op-amp compares the value of V_{IN} with a reference voltage which is generated by a chain of five 10 kΩ resistors. There is a 1 V drop across each resistor. So the bottom op-amp has a reference voltage of $+1$ V, the next one up has a reference voltage of $+2$ V, etc.. The whole system is simply four of the circuits shown in figure 14.4 put together.

V_{IN} / V	D	C	B	A
<1	off	off	off	off
1 – 2	on	off	off	off
2 – 3	on.	on	off	off
3 – 4	on	on	on	off
>4	on	on	on	on

The system is called a **bargraph voltmeter**. Its output is not very precise, but it does give a very rapid visual indication of voltage. It can respond much more rapidly than a moving-coil voltmeter, but is much simpler and cheaper than an oscilloscope. It also is comparatively rugged. You have probably seen bargraph voltmeters used in music centres to tell the user how loud the music is.

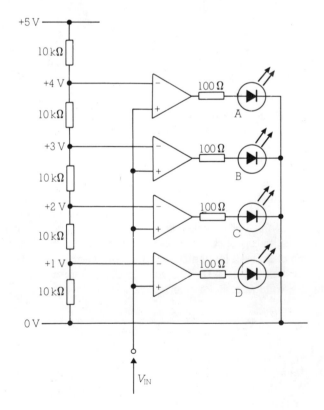

Figure 14.10 A bargraph voltmeter

QUESTIONS

1 Copy and complete the following statements.
  a) When the inverting input of an op-amp is at a higher voltage than the non-inverting input, the output is ... V. When the inverting input is at a lower voltage than the non-inverting input, the output is ....
  b) The input of an op-amp with the − sign next to it is called .... The input with the + sign is called .....
  ....
  c) An op-amp is usually run off .... supply rails of ... and .....

2 This question is about the circuit of figure 14.11.
  a) Calculate the voltage at R.
  b) What is the voltage at S when V_{IN} is
     i) +4 V,
     ii) +2 V,
     iii) 0 V,
     iv) −2 V?
  c) The system could be a voltage indicator for the blind. What range of values of V_{IN} will make the buzzer work?
  d) Adapt the circuit of figure 14.11 so that the buzzer makes a noise if the light gets too dim. Provide some means for adjusting the light level at which the buzzer goes off. Explain how your system works.

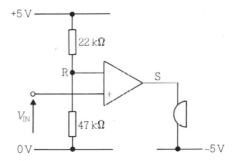

Figure 14.11 Question 2

3 Calculate the output voltages of the voltage dividers shown in figure 14.12.

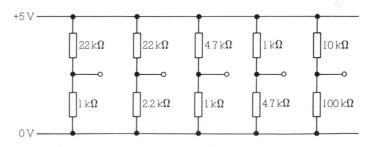

Figure 14.12 Question 3

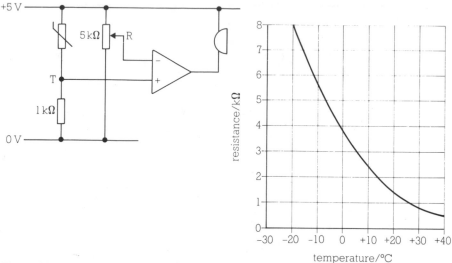

Figure 14.13 Question 4

4 The circuit of figure 14.13 is supposed to be a frost alarm. If the temperature of the thermistor goes below 0 °C, the buzzer makes a noise.

    a) Use the graph of figure 14.13 to read off the resistance of the thermistor at
       i)   + 10 °C,
       ii)   0 °C,
       iii) − 10 °C.

    b) Calculate the voltage at T when the thermistor is at 0 °C.

    c) What voltage should R be set at for the circuit to work properly?

    d) What happens to the voltage at T as the thermistor is cooled from + 10 °C to − 10 °C? What does the buzzer do while this is going on?

5 Work out the output voltages of the two op-amp circuits shown in figure 14.14.

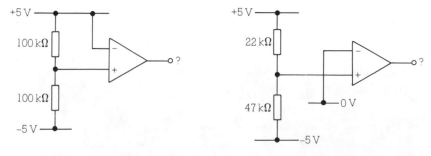

Figure 14.14 Question 5

6 Describe the difference between a digital signal and an analogue one.

7 Over what range of values of V_{IN} will the LED of figure 14.15 glow? Explain your answer carefully. (Hint; calculate the voltages at X and Y first.)

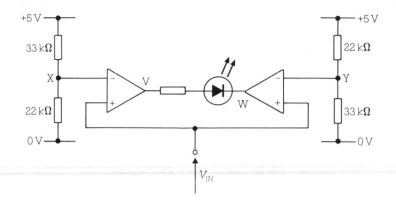

Figure 14.15 Question 7

15
Audio systems

This chapter is going to show you how electronics can be used to process **sound**. Most of the circuitry will be discussed in block diagram form. The next two chapters will show you how those blocks can be fleshed out with op-amps and transistors.

Audio signals

An audio signal is what comes out of a **microphone** when a sound wave is fed into it. Alternatively, it is what you have to feed into a **speaker** in order to create a sound.

Figure 15.1 shows the CRO trace of a typical audio signal. It is a **sine wave** which has a frequency of about 330 Hz. Like all audio waveforms it **alternates**, spending half of its time at a negative voltage and the rest of the time at a positive voltage. Audio signals are **AC** signals, where AC stands for alternating current.

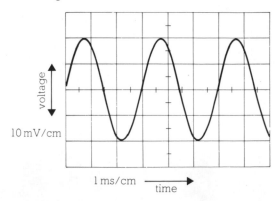

Figure 15.1 A typical audio signal

Microphones

The input transducer which converts sound into an electrical signal is called a **microphone**. Its circuit symbol is shown in figure 15.2. The trace of figure 15.1 is a typical microphone signal; it is what you would get if you held a 330 Hz tuning fork in front of it. Note how small the signal is; it has an amplitude of only 20 millivolts (**1000 mV = 1 V**).

Human beings can hear sounds whose frequencies lie between 16 Hz and 16 kHz. There are several types of microphone, each with their

Figure 15.2 A microphone

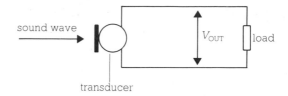

Figure 15.3 A microphone connected to a load

own characteristics, which can detect this full range. The better ones are more expensive. **Crystal microphones** are cheap and have a relatively large output. **Moving coil** types produce better waveforms but have a smaller output signal and are not very robust. All of our examples are going to assume the use of a crystal microphone.

Crystal microphones have a high **output impedance**. This means that they only function properly if they are feeding their signal into a high resistance **load**. If the load of figure 15.3 is much less than 100 kΩ, the amplitude of the microphone's signal will be considerably reduced.

Speakers

Microphones absorb energy from sound waves. That energy is used to make alternating currents flow through the load to which the microphone is connected (figure 15.3). In particular, if the load is a speaker (figure 15.4) the electrical energy can be converted back into sound energy. Microphones and speakers allow sound to be transmitted from one place to another via a pair of wires.

Figure 15.4 Transmitting sound along wires

Speakers tend to have very low resistances, typically 8 Ω. To make a reasonable sound they need to be fed with energy at a rate of at least 100 mW. (A typical power for a domestic speaker is 10 W.) So you need to make a lot of current go through a speaker (at least 100 mA) before it makes a sound which can be easily heard.

Making it bigger

The circuit of figure 15.4 does not work. The microphone does not provide enough electrical energy to allow the speaker to operate effectively. The signal coming out of the microphone needs boosting (or amplifying) before it is fed into the speaker. Devices which can boost AC signals are called **amplifiers**.

Amplifiers

The general circuit symbol for an amplifier is shown in figure 15.5. Its function is to create an output waveform which is exactly the same as its input waveform, except that it has twice the amplitude. It has a **voltage gain** of two.

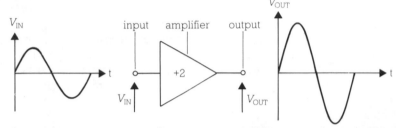

Figure 15.5 An amplifier with a gain of two

The output waveform of a good amplifier will always have the same shape as its input waveform. If the two shapes are not the same, the amplifier is **distorting** the signal. This is shown in figure 15.6. A distorted waveform will make the wrong sound when it is fed into a speaker. Distortion is obviously a bad thing!

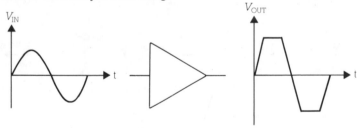

Figure 15.6 An amplifier with a distorted output

Measuring gain

To measure the gain of an amplifier you have to feed a **test signal** into it. A device called a **signal generator** is usually used to feed a small AC signal into the amplifier. A 10 mV amplitude sine wave with a frequency of 300 Hz might be suitable. A CRO can then be used to measure the amplitude of the signal fed out of the amplifier. (It can also be used to check for any distortion of the output waveform.) A suitable arrangement is shown in figure 15.7.

For example, suppose that the amplifier feeds out a 200 mV amplitude signal when a 10 mV signal is fed in. The gain will be given by this formula.

$$\text{Gain} = \frac{\text{output signal}}{\text{input signal}} = \frac{200}{10} = 20$$

Saturation

The amplifier should have a gain which is independent of the amplitude of the signal being amplified. So a 1mV waveform fed in causes a similar 20 mV waveform to be fed out. However, if the input signal gets too large the amplifier will **saturate** and distort the signal, as shown in figure 15.6. This is because the amplifier's supply rails put a limit on the largest output signal that it can generate. The amplifier of figure 15.7 is run off supply rails of +5 V and −5 V so it could not generate a waveform with an amplitude of more than 5 V.

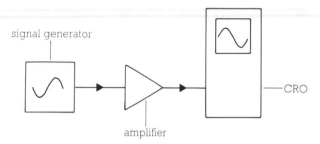

Figure 15.7 Testing an amplifier's gain

Power gain

For practical reasons two types of amplifier need to be employed between a sound source and a speaker. One amplifier provides **voltage gain**. The other provides **power gain**. This is shown figure 15.8.

A **power amplifier** has a voltage gain of only 1, so its output waveform is exactly the same as its input waveform. Its output terminal, however, must be able to provide the large current needed to run the speaker attached to it. That current, of course, comes from the supply rails.

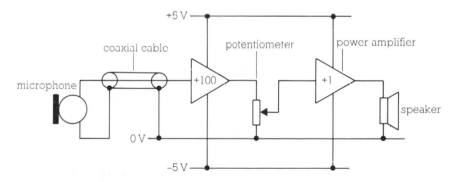

Figure 15.8 A typical audio system

Volume control

The system of figure 15.8 contains an amplifier with a fixed gain of 100. The overall gain that you want the system to have will depend on the strength of the sound being fed into the microphone. So a potentiometer is provided to allow some control over the amplitude of the signal being fed into the power amplifier. It acts as a variable voltage divider. The overall gain of the system can therefore be set at any value betwen 100 and 0 by rotating the potentiometer knob. A high gain, of course, means that a loud sound will be produced by the speaker.

Interference

Figure 15.8 shows the microphone connected to the input of the amplifier by a length of **coaxial cable**. The outer sheath of the cable is held at 0 V (known as **ground**) and the central conductor carries the alternating voltage from the microphone to the amplifier. This arrangement stops the small signal from the microphone being swamped by **mains hum**. The source of this interference is mains electricity. It produces alternating magnetic fields which can make AC signals at 50 Hz appear in wires. Since all input transducers for audio signals produce small AC signals it is important that they are connected to amplifiers by correctly grounded coaxial cables.

Storing sound

Two methods are widely used for storing audio signals so that they can be used to create sound some time in the future. They employ **discs** and **magnetic tape**.

Disc storage

Disc storage is the oldest technique. The audio signal is stored on the surface of a flat plastic disc in the form of a long wavy groove. The groove is a continuous one, spiralling from the outside of the disc to the inside. The side to side wiggles of the groove are like the trace seen on a CRO screen. The sideways motion of the groove mirrors the change in voltage of the audio signal being stored.

A disc playback system uses a needle placed in the groove to extract the audio signal from the disc. As the disc is rotated at a constant speed the needle moves from side to side with the groove. That motion can be used to recreate the audio signal via a piezoelectric crystal or a magnetised coil.

Tape storage

Magnetic tape stores sound as a variation in the magnetisation of a thin layer of magnetic material spread on one side of a long thin plastic tape. To record the sound on tape the audio signal is fed into a small electro-magnet. As the tape is fed past the electromagnet at a steady speed its magnetisation is altered. So the pattern of magnetisation along the tape mirrors the variation in voltage of the audio signal which is recorded on it.

Noise

Both of the recording media (disc and tape) will saturate and distort the audio signal if it is too large. On the other hand, if the audio signal is too small, the noise present in the recording medium may drown it out. This **noise** is a random audio signal, due to imperfections of the recording medium. It is the hiss that you hear if you listen to a blank cassette tape with the volume turned up.

QUESTIONS

1 Copy and complete the following statements.
  a) A microphone converts the energy of a sound wave into an
     .... current. Sound waves can be produced by feeding
     an alternating current into a ....
  b) A crystal microphone produces a signal of approximate amplitude
     ..... If that signal is fed into a resistance of less than about ......
     it is seriously reduced. A speaker has a typical resistance of about
     ..... and needs at least ...... of power to make an audible
     sound.
  c) Human beings can detect sound whose frequency lies between
     ...... and ....
  d) An .... is a device which makes enlarged copies of AC
     signals. The gain of the device is ....
  e) Microphones are connected to amplifiers by .... to prevent
     interference. The main source of interference is ....

2 The following data is supplied with an integrated circuit amplifier, the
  MB500.

Parameter	Maximum	Typical	Minimum
Supply voltage	18 V	15 V	10 V
Input signal	500 mV	—	—
Operating temperature	100 °C	—	−40 °C
Output current	5 mA	—	—
Voltage gain	50	40	30
Input resistance	—	500 kΩ	—
Maximum frequency	10 kHz	8 kHz	7 kHz

  a) Is the device a voltage amplifier or a power amplifier? Give your
     reasons.
  b) Explain how you would go about checking the data given for the
     gain of the MB500.
  c) If you fed a 25 mV amplitude signal into the MB500, what is the
     typical amplitude of signal which would be fed out?
  d) Is the IC suitable for amplifying the signal from a crystal micro-
     phone? Explain your answer.
  e) In what respect is the IC far from ideal as a good amplifier of
     audio signals?
  f) What is likely to happen if you try to run the IC off a pair of supply
     rails that are 20 V apart?
  g) Estimate the largest amplitude signal you can feed into the ampli-
     fier without it being distorted when it comes out of the output.

16
Voltage amplifiers

This chapter is all about how an op-amp behaves when you apply **negative feedback** to it. It converts op-amps into very good audio amplifiers.

Followers

The simplest type of negative feedback for an op-amp is shown in figure 16.1. The output of the op-amp has been connected directly to its inverting input. That feedback path forces the op-amp to have the transfer characteristic shown in figure 16.1.

The output voltage is the same as the input voltage.

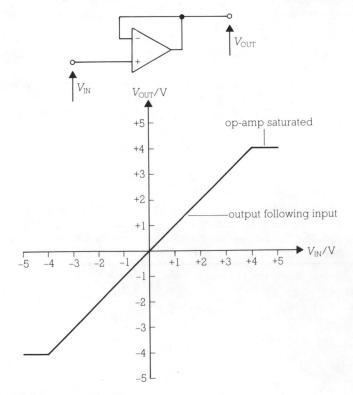

Figure 16.1 An op-amp follower

Of course, the output of the op-amp cannot rise above +4 V or fall below −4 V. So if V_{IN} goes too high or too low the op-amp saturates and the output voltage no longer follows the input voltage.

Negative feedback

The follower shown in figure 16.1 is an amplifier with a gain of +1. Yet an op-amp on its own behaves a bit like a logic gate. Its output has one of two states (+4 V or −4 V) depending on which of its inputs is held at the higher voltage. Why does negative feedback convert it into an amplifier?

Differential amplifiers

In fact, an op-amp is always an amplifier. It feeds out a signal which is about 100 000 times larger than the signal which is fed into it. That input signal is the **difference** in voltage of the input terminals. So an op-amp measures the difference in voltage of its two input terminals, multiplies the result by 100 000 and feeds it out of the output. Of course, the op-amp is going to saturate unless the voltage difference of the inputs is less than 0.004 mV.

The infinite gain approximation

When you apply negative feedback to an op-amp you are allowing its output to adjust the voltage of the inverting input. (See figure 16.1.) If the op-amp is not saturated then its two inputs must have almost the same voltage (certainly within 0.004 mV). The difference is so small that we can assume that it is zero. This is called the **infinite gain approximation**.

When negative feedback is applied to an op-amp
the output voltage adjusts itself so
that the two inputs have the same
voltage as each other.

This rule can be used to explain the operation of **any** op-amp which has negative feedback applied to it.

Non-inverting amplifiers

If you feed the output of an op-amp directly into its inverting input you end up with an amplifier whose gain is +1. Higher gains can be achieved by feeding the output back to the input via a voltage divider.

This is shown in figure 16.2. The two resistors feed one third of the

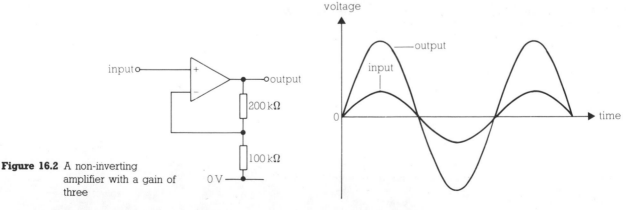

Figure 16.2 A non-inverting amplifier with a gain of three

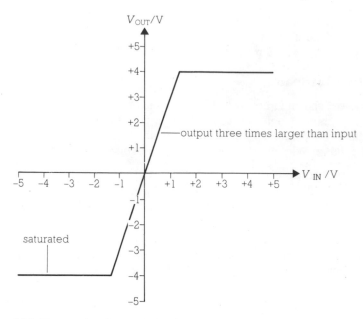

Figure 16.3 The transfer characteristic of the circuit in figure 16.2

output voltage back to the inverting input. Since both inputs will have the same voltage, this means that the output will always be three times larger than the input. This is shown in the transfer characteristic of figure 16.3.

As the graph of figure 16.2 shows, the system has a voltage gain of $+3$. The output waveform has an amplitude which is three times greater than that of the input waveform. The gain is positive because the input and output voltages always have the same sign as each other.

Calculating gain

The gain of a non-inverting amplifier is fixed by the the two resistors used in the voltage divider which connects the output to the inverting input. It can be calculated with this formula. (The symbols are defined in figure 16.4.)

$$G = \frac{V_{\text{OUT}}}{V_{\text{IN}}} = \frac{R_{\text{T}} + R_{\text{B}}}{R_{\text{B}}}$$

It is not difficult to see where this formula comes from. We can use the voltage divider formula to work out what V_{F} is.

$$V_{\text{F}} = \frac{R_{\text{B}} \times V_{\text{OUT}}}{R_{\text{B}} + R_{\text{T}}}$$

But the infinite gain approximation says that $V_{\text{F}} = V_{\text{IN}}$. So we get this formula.

$$V_{\text{IN}} = \frac{R_{\text{B}} \times V_{\text{OUT}}}{R_{\text{B}} + R_{\text{T}}}$$

If you move the symbols around, the last expression becomes the same as the gain formula quoted above.

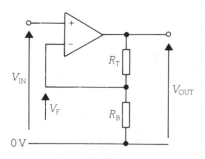

Figure 16.4 Symbols used in the gain formula

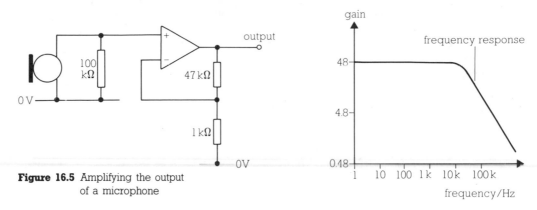

Figure 16.5 Amplifying the output of a microphone

Figure 16.6 The frequency response of an amplifier

Amplifying audio signals

Figure 16.5 shows a non-inverting amplifier in action. Its gain is easily calculated.

$$G = \frac{R_T + R_B}{R_B} = \frac{47 + 1}{1} = 48$$

So if the microphone feeds out a 2 mV signal the amplifier feeds out a 48 × 2 = 96 mV signal. Note how the microphone feeds its signal into a 100 kΩ load; you must always tie the non-inverting input of the op-amp to 0 V with a resistor for it to work properly.

Bandwidth

A single op-amp cannot give you a gain of more than about 50 if you want to amplify audio signals without distortion. The graph of figure 16.6 shows the **frequency response** of the amplifier in figure 16.5. It shows how the gain of the system depends on the frequency of the signal which it is amplifying.

The gain is a constant 48 for frequencies below about 20 kHz. Above 20 kHz the gain drops. The **bandwidth** of the system is 20 kHz; this tells you the range of frequencies that it will effectively amplify. The gain of an amplifier is linked with its bandwidth. **If you increase its gain you decrease its bandwidth**. So if you built a similar amplifier with a gain of 200, it would have a bandwidth of only about 5 kHz. As it would not be able to amplify the whole audio range of 16 Hz to 16 kHz, it would not make a very good audio amplifier.

AC coupling

To get a gain of more than 50 you have to connect amplifiers in series with each other. This is shown in figure 16.7. Each amplifier is of the type

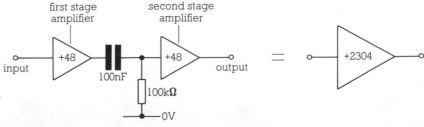

Figure 16.7 Coupling amplifiers to increase the gain

shown in figure 16.5. The output of the **first stage** amplifier is fed into the input of the **second stage** amplifier via a 100 nF capacitor. The whole system has a gain of 48 × 48 = 2304.

The amplifiers have been **AC coupled**. This means that only AC signals can get from the output of the first stage into the input of the second stage. Any DC signals which are fed out of the first stage are blocked by the **coupling capacitor**. (A DC signal is one which has a constant voltage.) High gain amplifiers generally feed out small DC signals which could create havoc if they were amplified further. The use of coupling capacitors between stages prevents this happening.

The coupling capacitor affects the frequency response of the whole amplifier. As you can see from the graph of figure 16.8, the gain starts to fall below 2300 when the frequency goes below 10 Hz. A 1 μF capacitor in place of the 100 nF (0.1 μF) would have allowed signals down to 1 Hz through without attenuation.

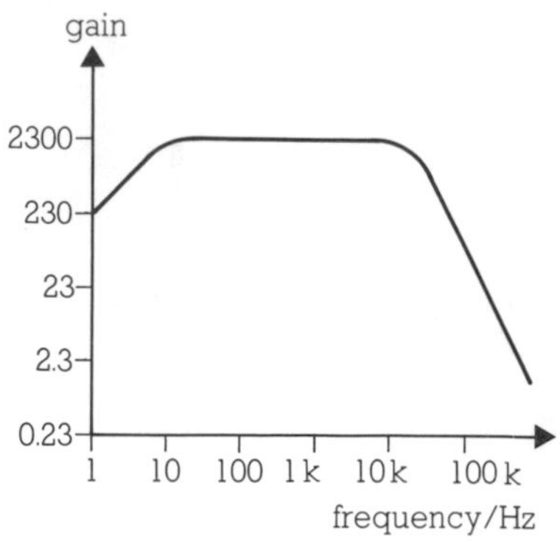

Figure 16.8 Frequency response of the circuit of figure 16.7

Stability

Although the circuit of figure 16.7 has a large gain it is not a very good audio amplifier. This is because it lacks **stability**; it is likely to oscillate rather than amplify!

It is very difficult to prevent large AC signals causing smaller AC signals in nearby wires. So unless you assembled the circuit of figure 16.7 very carefully (spacing out the components, keeping wires short and straight) the output could easily feed a small stray signal back into the input. The gain is very large (about 2300), so a much enlarged copy of the stray signal would instantly appear at the output. Some of that would stray back into the input etc., etc.

The secret of avoiding an unstable amplifier is to make its overall gain negative. The system must contain an inverting amplifier.

Inverting amplifiers

Figure 16.9 shows how an op-amp can be made into an **inverting amplifier** with a gain of −3. The gain is negative because (as the graph shows) the input and output voltages always have different signs. So when V_{IN} is positive V_{OUT} will be negative and vice versa.

The gain formula

The gain of an inverting amplifier is given by this formula. The symbols are defined in figure 16.10.

$$G = \frac{V_{OUT}}{V_{IN}} = \frac{-R_F}{R_{IN}}$$

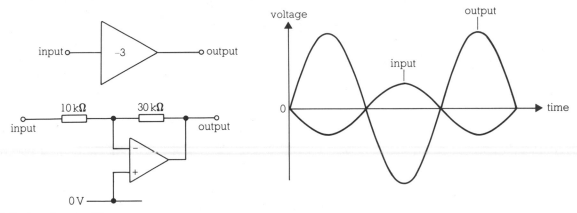

Figure 16.9 An inverting amplifier
with a gain of three

It is quite easy to see where the formula comes from. In figure 16.10 the output is fed back to the inverting input via R_F. So the feedback is negative. Both inputs of the op-amp must therefore have the same voltage. So the point marked '**virtual earth**' in figure 16.10 will be at 0 V.

Now we can use Ohm's law to work out how much current goes through R_{IN} (see figure 16.11).

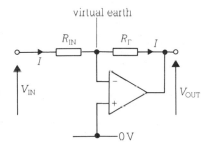

Figure 16.10 Symbols used in the
gain formula

$$R = \frac{V}{I} \qquad \text{therefore } R_{IN} = \frac{V_{IN} - 0}{I}$$

$$\text{therefore } I \quad = \frac{V_{IN}}{R_{IN}}$$

Figure 16.12 shows how the current I flows through the feedback resistor R_F. This is because the inputs of an op-amp have a very high input resistance. They don't let much current go into them. Ohm's law can be used to work out the voltage across R_F due to I.

$$R = \frac{V}{I} \qquad \text{therefore } R_F = \frac{0 - V_{OUT}}{I}$$

$$\text{therefore } V_{OUT} = -I \times R_F$$

$$\text{therefore } V_{OUT} = - \frac{V_{IN}}{R_{IN}} \times R_F$$

The last expression, after some shuffling around, becomes the gain formula.

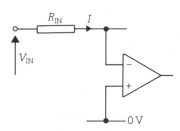

Figure 16.11 Current flowing into
the amplifier

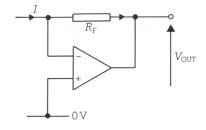

Figure 16.12 Current going through
the feedback resistor

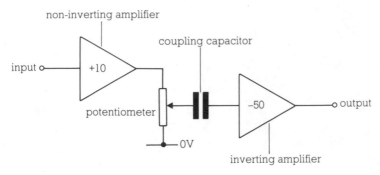

Figure 16.13 An amplifier with a gain variable from 0 to −500

Variable gain

A complete amplifier system, is shown in figure 16.13. The first stage uses the non-inverting amplifier of figure 16.14. It has a fairly low gain (+10) to keep it stable, but it has a very high input resistance (220 kΩ). So it will draw very little current from signal sources (such as microphones) connected to its input.

The second stage uses an inverting amplifier of the type shown in figure 16.15. Its gain can be easily calculated.

$$G = \frac{-R_F}{R_{IN}} = \frac{-500}{10} = -50$$

The inverting amplifier has a low input resistance (10 kΩ) so it is unsuitable as the first stage amplifier. It can, however, have a high gain because it is more stable than the non-inverting amplifier.

The whole system of figure 16.13 has a gain which can be varied from 0 to +10 × −50 = −500 by rotating the potentiometer knob.

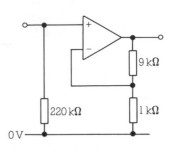

Figure 16.14 The first stage amplifier

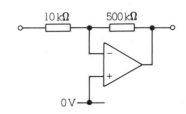

Figure 16.15 The second stage amplifier

QUESTIONS

1 Figure 16.16 is the block diagram of an amplifier system.
  a) Draw circuit diagrams for each of the blocks. Show suitable component values each time.
  b) Draw a circuit diagram for the whole system.
  c) State the value of its
    i) maximum gain,
    ii) minimum gain.

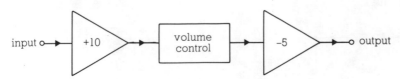

Figure 16.16 Question 1

2 Explain what is meant by the following terms as applied to amplifiers.
  a) Gain.
  b) Bandwidth.
  c) Stability.

3 Figure 16.17 shows a number of op-amp amplifiers. For each one, state
   a) if it is inverting or non-inverting,
   b) the value of its gain,
   c) its output voltage.

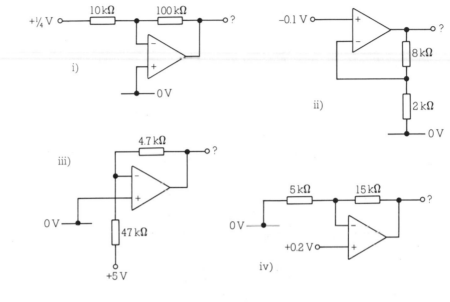

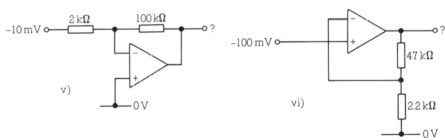

Figure 16.17 Question 2

4 Draw circuit diagrams to show how op-amps can be made into amplifiers with gains of
   a) + 11,
   b) − 1,
   c) − 0.1,
   d) + 5.
   Make the smaller of the two resistors 1 kΩ each time.

5 You are going to design an amplifier system with the following properties. Its gain must be variable, from 0 to 200. It must be stable. Its first stage must have an input resistance of 47 kΩ.
   a) Draw a block diagram of the whole system. Explain the function of each block.
   b) Draw a circuit diagram for the whole system.

6 A signal generator produces the waveform shown in figure 16.18. That signal is fed into each of the two amplifiers of figure 16.18. Copy the input waveform and, using the same axes, carefully draw the output waveform for each amplifier.

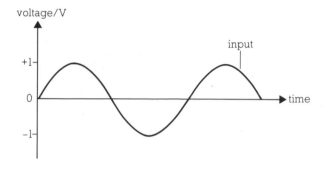

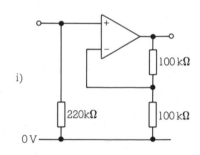

i)

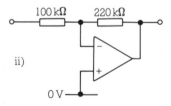

ii)

Figure 16.18 Question 6

17
Power amplifiers

Power amplifiers are devices which can drive large AC currents through low resistance output transducers. They usually have a voltage gain of + 1 and draw very little current from their signal source.

Op-amp followers

An op-amp wired up as a follower (figure 17.1) makes a simple (but generally useless) power amplifier. Very little current has to flow into its input (much less than a microamp), it has a voltage gain of + 1 and it can drive currents of up to about 10 mA through loads attached to its output. This may be enough to light up an LED, but it is far too small to make a speaker work effectively.

Power gain

Figure 17.2 shows the answer if you need to drive a speaker. A **transistor** power amplifier has to be inserted between the op-amp output and the speaker. The transistors arrange for lots of current to be delivered from the supply rails into the speaker. The feedback ensures that the voltage gain of the whole system is still + 1. So the waveform fed into the input of the system exactly matches the waveform being fed into the speaker. This means that the power amplification introduces no distortion.

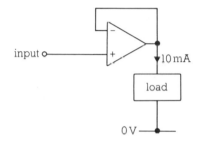

Figure 17.1 An op-amp follower driving a small load

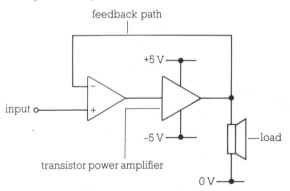

Figure 17.2 Using a power amplifier inside the feedback loop

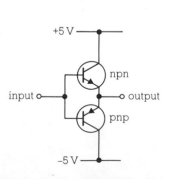

Figure 17.3 A push-pull follower

Generating heat

The transistor power amplifier that we shall be using is shown in figure 17.3. It is called a **push-pull follower** because of the way that the transistors work. (It is not the only way in which transistors can be used to build a power amplifier, but it is the simplest and functions well when inserted into the circuit of figure 17.2.)

Power amplifiers are usually made of discrete components. A power amplifier tends to generate a lot of heat, so it needs to be fairly bulky to prevent its temperature getting too high. It is one area of electronics where miniaturisation is not usually possible. Low power systems can be made into ICs if they are carefully cooled, but high power systems have to be made with individual transistors.

Transistors

Two types of transistor are used in the system of figure 17.3. The **npn** one directs current from the + 5 V supply rail into the output. The **pnp** controls the current from the output into the − 5 V supply rail.

npn transistors

The circuit symbol for an npn transistor is shown in figure 17.4. Each of its terminals has a name. The **base** is used to control how much current goes from the **collector** to the **emitter**.

Figure 17.5 shows an npn transistor in action. It is being used to control the current going through a light bulb. The emitter acts as the top supply rail for the bulb. Its voltage is fixed by the voltage at the base according to this approximate rule.

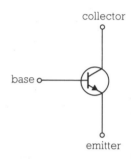

Figure 17.4 An npn transistor

The emitter is 0.7 V below the base.

So if the base is held at + 2.7 V by the potentiometer the emitter will be at 2.7 − 0.7 = +2.0 V. This makes a current of 200 mA go through the bulb. Most of that current (198 mA) comes from the + 5 V supply rail via the collector. A small amount (2 mA) comes from the base.

On and off

A transistor is **on** when current flows through it. This happens when the base is 0.7 V above the emitter. Should the base not be 0.7 V above the emitter, the transistor will switch **off**. No current will flow in or out of any of the three terminals.

Current gain

Transistors are useful because they provide current gain. A small current flowing into the base of an npn transistor controls a much larger current flowing from collector to emitter.

The **current gain** of a transistor is given (for historical reasons) the symbol **h$_{FE}$**.

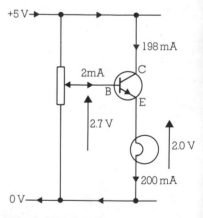

Figure 17.5 Using an npn transistor to control the voltage across a bulb

$$I_C = h_{FE}\, I_B$$ I_C is the collector current,
I_B is the base current,
h$_{FE}$ is the current gain.

Typically h$_{FE}$ has a value between 50 and 400. Some types of transistors have higher current gains than others, but even among transistors of the same type there is quite a lot of variation. We are going to assume that h$_{FE}$ is 100 in our examples. Since the collector current is so much larger than the base current, it is a good approximation to say that the emitter and collector currents are the same.

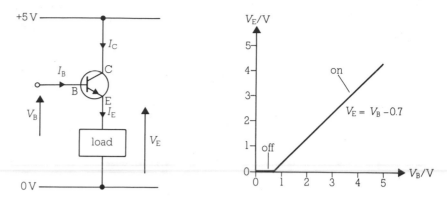

Figure 17.6 How the base voltage is related to the emitter voltage

The npn rules

The approximate behaviour of an npn transistor can be summarised in the rules stated below. The symbols used are defined in figure 17.6.

The transistor is switched off if V_B is not 0.7 V above V_E.
When the transistor is switched on V_B is 0.7 V above V_E.
$$I_C = h_{FE} I_B \simeq I_E$$

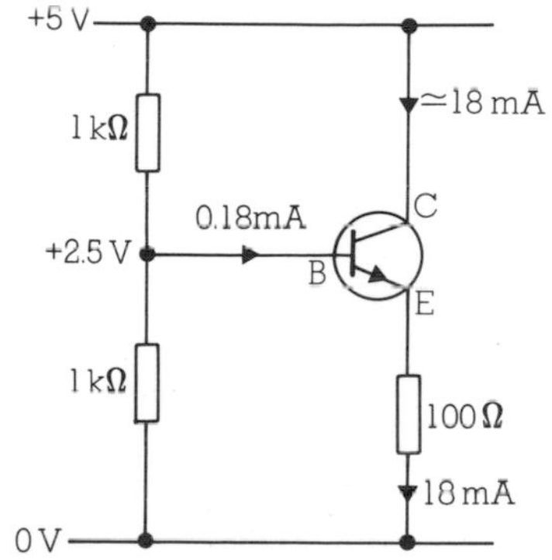

As an example of how these rules are used, look at figure 17.7. We want to calculate how much current the transistor will draw from the top supply rail.

The base is held at +2.5 V by the pair of 1 kΩ resistors arranged as a voltage divider. So the emitter will be at +2.5 − 0.7 = +1.8 V.

Ohm's law can now be used to calculate how much current goes through the 100Ω resistor.

$$R = \frac{V}{I} \qquad \begin{array}{l} R = 0.1 \text{ k}\Omega \\ V = 1.8 \text{ V} \\ I = ? \end{array} \qquad \text{therefore } 0.1 = \frac{1.8}{I}$$

$$\text{therefore } I = \frac{1.8}{0.1} = 18 \text{ mA}$$

Figure 17.7 Currents and voltages at the three terminals of an npn transistor

The current through the 100Ω resistor is the same as the emitter current. As the collector current is virtually the same as the emitter current, a current of 18 mA will be drawn from the top supply rail by the transistor. If the current gain of the transistor is 100, the current drawn from the voltage divider by the base will be only 0.18 mA.

pnp transistors

The npn transistor of figure 17.7 is being used to **source** current into the 100 Ω resistor. Its emitter is acting as the top supply rail for the resistor.

If you want to **sink** current from a load you need to use a **pnp** transistor. This is shown in figure 17.8. The emitter of the transistor is acting as the bottom supply rail for the bulb.

Notice how the pnp transistor differs in behaviour from the npn transistor. Its circuit symbol is shown in figure 17.9; note how the emitter arrow is different. Current goes into the emitter and comes out of the base and collector. Furthermore, you hold the base 0.7 V below the emitter to turn the transistor on. Like the npn transistor, the collector current of the pnp transistor is much larger than the base current.

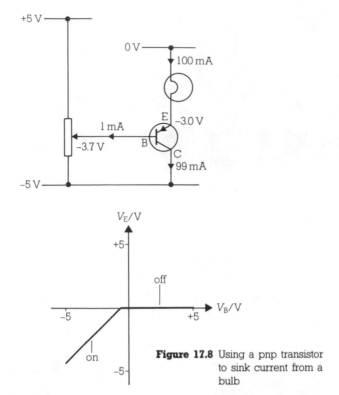

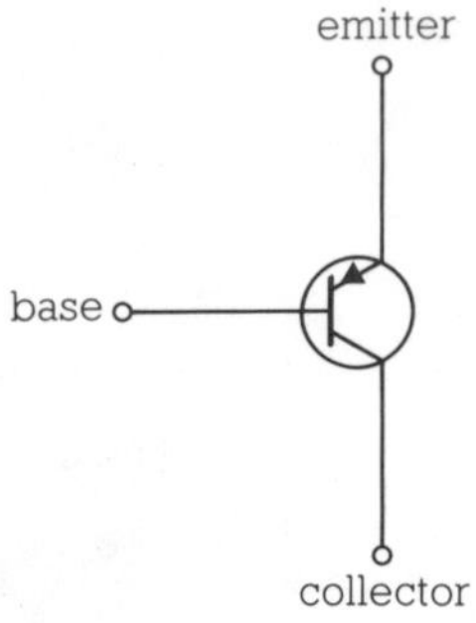

Figure 17.9 A pnp transistor

Figure 17.8 Using a pnp transistor to sink current from a bulb

The pnp rules

The approximate rules obeyed by pnp transistors are shown below.

The transistor is switched off when V_B is not 0.7 V below V_E
The transistor is switched on when V_B is 0.7 V below V_E
$$I_C = h_{FE} \, I_B \simeq I_E$$

The push-pull follower

A power amplifier for analogue signals has to be able to feed out both positive and negative voltages. When its output is positive it will have to source current into the load like an npn transistor does. When its output is negative it will have to sink current from the load like a pnp transistor does. Put both types of transistor together as shown in figure 17.10 and you have a system which can both source and sink current. It is called a **push-pull follower**.

The two examples of figure 17.11 show how the push-pull follower works.

When V_{IN} is positive the npn transistor turns on and sources current into the load. V_{OUT} will be 0.7 V below V_{IN}.

When V_{IN} is negative the pnp transistor turns on and sinks current from the load. V_{OUT} will be 0.7 V above V_{IN}.

Both transistors will be turned off if V_{IN} is between +0.7 V and −0.7 V. So V_{OUT} will be 0 V.

The behaviour of the system is summarised in the transfer characteristic of figure 17.10. **One transistor handles positive voltages, the other handles negative voltages.**

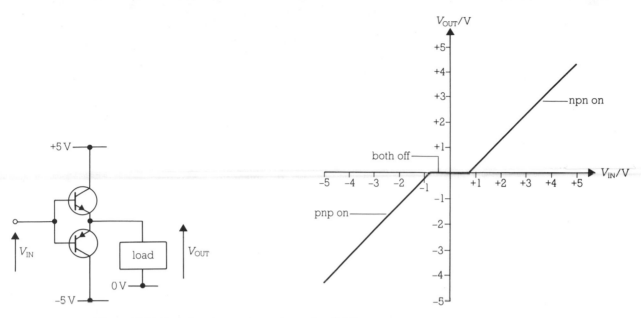

Figure 17.10 Transfer characteristic of a push-pull follower

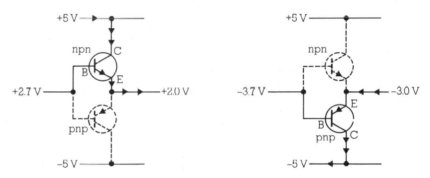

Figure 17.11 One transistor pushes, the other one pulls

Crossover distortion

The transfer characteristic of a push-pull follower is not a straight line. This means that it will distort signals that are fed into it.

For example, suppose that you used a push-pull follower to feed the signal from a signal generator into a speaker as shown in figure 17.12.

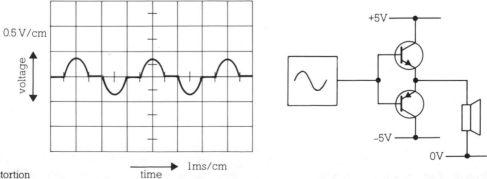

Figure 17.12 Crossover distortion

Although the signal generator feeds out a 1 V amplitude sine wave at 330 Hz, the signal fed into the speaker will look nothing like a sine wave. A typical CRO trace for the output waveform is shown in figure 17.12. It shows the **crossover distortion** quite clearly. The waveform is flat during the time that both transistors are switched off. Only when the input waveform is above $+0.7$ V or below -0.7 V will current flow through the speaker. So if the amplitude of the input waveform had been 0.5 V instead of 1 V, no sound at all would have been produced by the speaker!

Negative feedback

Crossover distortion can be eliminated by using an op-amp to control the transistor bases. Look at figure 17.13. V_{OUT} is forced to be the same as V_{IN} by the negative feedback applied to the op-amp. So the waveform fed into the speaker is not distorted. A typical CRO trace for that waveform is shown in figure 17.13; it is the same as the waveform fed out of the signal generator. The negative feedback has made the system **linear**.

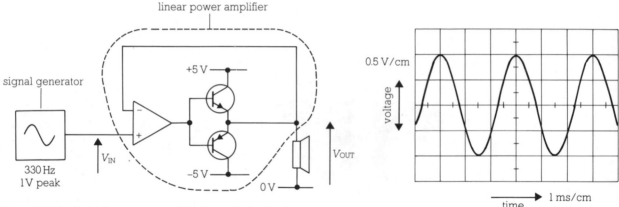

Figure 17.13 Eliminating crossover distortion with feedback

The transfer characteristic of the system is shown in figure 17.14. Note that it is a straight line until the op-amp saturates. So waveforms whose amplitude is less than 3.3 V will not be distorted by the system.

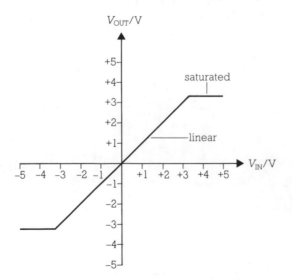

Figure 17.14 The transfer characteristic of a linear power amplifier

Transistor ratings

Many different types of transistor are available. How do you decide which one to use for a particular circuit?

For example, suppose that you wanted to choose a transistor for the sound-to-light converter shown in figure 17.15. There are three ratings to be considered for the transistor.

It must be able to cope with its collector being held 5 V above its emitter.

It must be able to survive a collector current of 300 mA (the current rating of the light bulb).

It must have a power rating of about 500 mW. Most of the heat in a transistor is generated by the collector current. When this is at its greatest (300 mA) the collector-emitter voltage will be $5 - 3.3 = 1.7$ V. So the power will be $VI = 1.7 \times 300 = 510$ mW. The average power will not be much different from this.

You can increase the power rating of a transistor by connecting it to a **heat sink**. This is a large finned piece of black metal which helps to get rid of the heat. It pays to let transistors run cool; if you let one get too hot it will self-destruct rapidly.

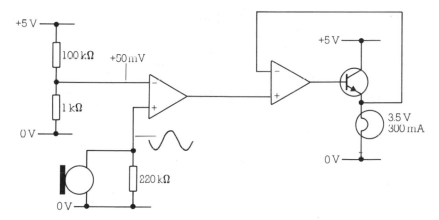

Figure 17.15 A sound-to-light converter

QUESTIONS

1 a) Draw the circuit symbol of an npn transistor. Label the base, collector and emitter. Show with arrows which way the current flows in each of those terminals.

  b) Copy and complete the following statements.
  No current flows in an npn transistor if the base is less than .... V above the ..... When current flows out of the emitter, most of it comes from the ..... For current to flow out of the emitter, the .... must be .... V above the ..... The .... current is slightly less than the emitter current. The .... current is much less than the emitter current.

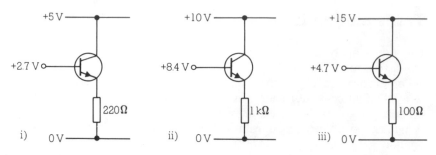

Figure 17.16 Question 2

2 This question is about the three circuits shown in figure 17.16. Each of the transistors has a current gain of 100. For each of the circuits, calculate
a) the voltage at the emitter,
b) the emitter current,
c) the collector current (roughly),
d) the base current.

3 a) Draw the circuit symbol of a pnp transistor. Label the base, collector and emitter. Show with arrows which way the current flows in each of those terminals.

b) Copy and complete the following statements.
No current flows in a pnp transistor if the base is not .... V below the emitter.
When current flows into the emitter, most of it flows out of the ..... For current to flow into the emitter the base must be .... V below the ..... The .... current is much smaller than the emitter current.

4 This question is about the three circuits shown in figure 17.17. Each transistor has a current gain of 200. For each of the circuits, calculate
a) the voltage at. the emitter,
b) the emitter current,
c) the collector current (roughly),
d) the base current.

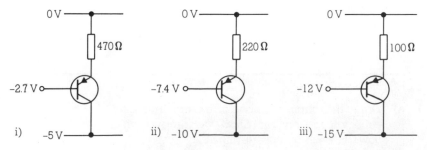

Figure 17.17 Question 4

5 Draw the circuit diagram of a linear power amplifier which uses a push-pull follower and an op-amp. Draw the transfer characteristic of the system. If the system is run off split supply rails of +9 V and −9 V, what is the maximum amplitude of signal which the system can deal with without adding distortion?

6 Figure 17.18 shows a push-pull follower controlling the current through a load. Complete this table for that circuit.

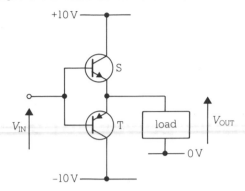

Figure 17.18 Question 6

V_{IN}/V	State of S (on/off)	State of T (on/off)	V_{OUT}/V
+8.7	?	?	?
+2.0	?	?	?
+0.5	?	?	?
−0.3	?	?	?
−3.3	?	?	?
5.7	?	?	?

7 You have to design an intercom system. A microphone has to detect your voice in one room, a speaker has to recreate your voice in another room. The system needs a volume control at the speaker end. The maximum gain required is 1000.
a) Draw a block diagram of the system using the following blocks. Speaker, microphone, volume control, power amplifier, inverting amplifier and non-inverting amplifier.
b) Draw a circuit diagram for each of the amplifier blocks, showing component values.

Using a heat sink to keep a component cool.

<h1 style="text-align:center">18</h1>

Simple power supplies

For the purposes of this chapter, a power supply is a device which uses mains electricity to generate a steady voltage difference between supply rails.

A five volt power supply is represented in figure 18.1. Three wires (L, N and E) lead into it from the mains electricity supply. Two wires (+5 V and 0 V) come out of it. A power supply like this might be used to run a logic system.

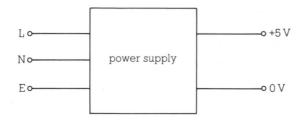

Figure 18.1 Block diagram of a power supply

Mains electricity

Mains electricity is lethal. Only people who are properly qualified should make mains-driven power supplies. You will **not** become properly qualified by simply studying this chapter.

The three lines

Mains electricity arrives in your home or the laboratory via a wall socket (figure 18.2). Each gap in it allows a suitable plug to be connected to the three **lines** of the mains supply.

Earth

The **earth line** is connected to the ground beneath the building containing the socket. It is at 0 V and is perfectly safe for you to touch. At some point in the building the earth line will be connected to a metal object which goes into the ground; the pipe which delivers water to the building is often used.

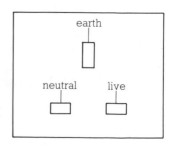

Figure 18.2 A mains electricity socket

Live

The **live line** carries an alternating current at a dangerously high voltage.

Figure 18.3 shows a graph of the voltage of the live line (V_L) as a function of time. It is a sine wave which has an amplitude of 340 V and a period of 20 ms. It is customary to quote the **root mean square** or **rms** voltage of the mains supply rather than its amplitude. The following rule is used to calculate the amplitude or **peak value** of an alternating voltage from its rms voltage.

$$V_{peak} = 1.414 \times V_{rms}$$

The live line carries a 240 V rms sine wave at 50 Hz.

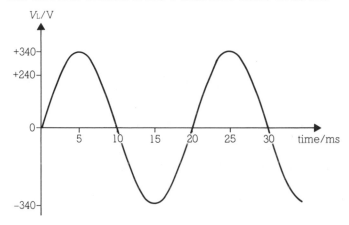

Figure 18.3 How the voltage of the live line varies with time

Neutral

The **neutral line** is usually at, or near 0 V. It is the second of the two wires needed to transport electrical energy from the local electricity substation to the wall socket. As you can see from figure 18.4, the neutral line is connected to the ground under the substation.

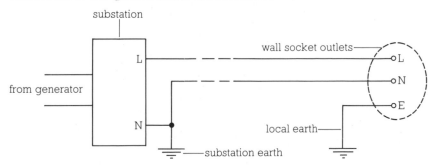

Figure 18.4 The three lines going into a mains socket

Mains connections

Devices which run off mains electricity are connected to wall sockets by **three pin plugs** and **three core cable**. The plug allows the three wires of the cable to be safely connected to the three lines in the socket.

Three core cable which is designed for mains connection contains three separate wires. Each has a colour coded insulation according to this internationally recognised code.

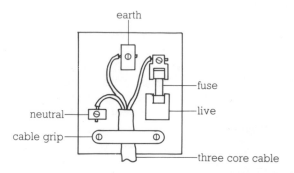

Figure 18.5 A mains plug

live	brown
neutral	blue
earth	green striped with yellow

As extra protection, a further layer of insulation surrounds all three wires and holds them together.

Figure 18.5 shows how the wires are connected to a three pin plug. **It is vital that a plug is always correctly and carefully wired up.** In particular, the screws should always be tightly screwed down onto the bared ends of the wire. The cable grip prevents the wires from being pulled out of place if the cable is tugged.

Mains fuses and safety

Figure 18.6 shows how an AC motor (or anything else) should be connected to the mains supply.

The earth wire must be firmly connected to all metal parts of the device's outer casing. This makes the system safe, as we will show below.

The live and neutral lines go to the motor terminals. A switch is included in the live line so that the device can be **isolated** from the live line. This not only turns it off, it also makes it safe to touch when it is not working.

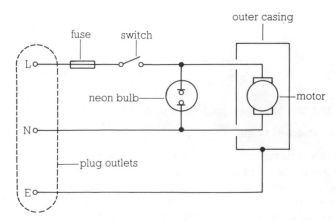

Figure 18.6 Connecting a motor to a mains plug

(A switch in the neutral line would also turn the device off, but it wouldn't isolate it. This would make the system dangerous.)

A **neon bulb** is placed between the live and neutral lines to indicate when the device is connected to the mains supply. It will glow red when the switch is closed. As neon bulbs only draw a few milliamps from the power supply they do not consume much energy. On the other hand, any power supply which can light up a neon bulb will certainly supply enough current to kill a human being.

Fuses

The **fuse** is a thin piece of wire enclosed in glass or ceramic which is inserted in the live line before the switch. It is usually in the three pin plug and acts as an automatic safety switch.

Each fuse has a **current rating**. If the current through it exceeds that rating the wire inside it will melt and break. The fuse will **blow**. So a fuse is a current sensitive switch. If too much current goes through the live line the fuse will blow and automatically switch that current off. Note that because it is in the live line, a blown fuse isolates the device, leaving it safe to touch.

There are three main ratings for mains fuses. They are 13 A, 5 A and 3 A. A 3 A fuse is suitable for most small power supplies.

Earth for safety

The live and neutral wires carry electrical energy from the wall socket to the motor of figure 18.6. The fuse and earth wire make the system safe. This is how they do it.

The top diagram of figure 18.7 shows how the current flows if all is well. (Remember that it is AC, so its direction will change one hundred times a second.) No current flows in the earth line.

The bottom diagram of figure 18.7 shows what happens when a dangerous fault develops. The insulation of the live wire has become worn through where it enters the metal casing. So the casing is touching the live wire and is therefore dangerous to touch. Fortunately, the casing is earthed. So current can get back to the substation via the ground instead of the neutral wire. If the earthing connections are sound, that route will have a low resistance. A very large current will flow in the live line, instantly blowing the fuse in it.

**So as soon as the fault develops the
live line is automatically cut off.**

Hazards of electricity

Direct contact with the live line of the mains supply is always unpleasant and often fatal.

It only requires a current of a few milliamps through the human body for its muscles to contract uncontrollably. Quite often a hand which is in contact with the live line will grasp that line firmly, making the shock worse. If the current crosses the chest, there is a strong chance that the heart will stop working. **Electricity can kill.**

If the system of figure 18.6 had not been earthed or fused, it would would have been unsafe. When the live wire touched the casing the motor would have still have continued to operate, so nothing would appear to be wrong. But if someone touched the casing its alternating voltage would make current go through her to the ground. The amount of current would

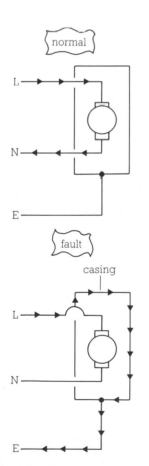

Figure 18.7 Current flow when a fault develops

depend on what the floor was made of, what type of shoes she was wearing and how wet she was (water would lower her resistance and therefore increase the current). It would probably be enough to kill her.

**So never tamper with the earth connections
or fuses of a power supply. They could
save your life one day.**

Transformers

The first stage in most power supplies is a **transformer**. It reduces the dangerous 240 V rms mains waveform to a lower and safer voltage.

A transformer contains two coils of wire wound on an iron core. An alternating voltage placed across the ends of the **primary coil** will cause a similar alternating voltage to appear across the ends of the **secondary coil**. The relative amplitudes of the two waveforms are given by this rule.

$$\frac{\text{primary voltage}}{\text{secondary voltage}} = \frac{\text{turns in primary coil}}{\text{turns in secondary coil}}$$

Good transformers are able to change the amplitude of an AC waveform without taking much energy out of it. So electrical energy passes from the primary to the secondary without generating much waste heat in the transformer.

AC to AC

Figure 18.8 shows how to wire up a transformer to the mains supply. The live and neutral lines go to the ends of the primary coil (P), feeding it with a 240 V rms sine wave at 50 Hz. A 9 V rms sine wave at 50 Hz appears across the ends of the secondary coil (S). The iron core of the transformer has been earthed to make it safe.

The transformer has a **power rating** of 12 W. You can feed electrical energy through it at a rate of 12 W without any danger of overheating it. So there is a limit to how much current you can draw from its secondary coil without melting its insulation.

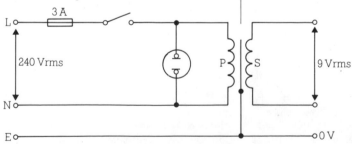

Figure 18.8 Connecting a transformer to the mains supply

$$W = VI \qquad W = 12 \text{ W} \qquad \text{therefore } 12 = 9 \times I$$
$$V = 9 \text{ V rms}$$
$$I = ? \qquad\qquad I = \frac{12}{9} = 1.3 \text{ A rms}$$

So any power supply which was built with this transformer would not be able to deliver a current of much more than 1 A safely.

The 9 V rms output of the transformer is quite safe to touch. The peak voltage across the secondary coil will be $1.414 \times 9 = 12.7$ V, not enough to cause any damage to human beings.

Rectification and smoothing

There are two stages necessary for the conversion of an AC signal into a DC one. They are called **rectification** and **smoothing**. Figure 18.9 illustrates what a rectifier and smoother does. It converts an alternating voltage (V_{AC}) into one which has a constant value (V_{DC}).

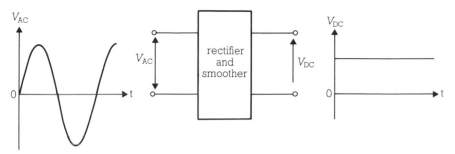

Figure 18.9 A rectifier and smoother

Diodes

The rectification is done with **diodes**. These act as one-way valves for electric current.

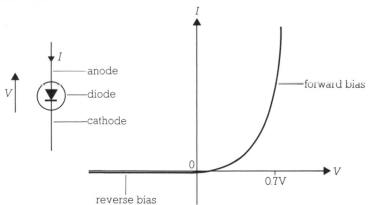

Figure 18.10 I–V characteristic of a diode

Figure 18.10 shows the circuit symbol and typical I-V characteristic of a silicon diode. Although they come in many shapes and sizes, all diodes have the same behaviour. Current flows easily from anode to cathode but finds it difficult to flow from cathode to anode.

The graph of figure 18.10 shows that once the anode of a silicon diode is more than 0.7 V above its cathode the current going through it increases very rapidly. When the diode is **forward biased** like this it has a low resistance. On the other hand, if you **reverse bias** it by holding its anode at a lower voltage than its cathode, the diode has a very high resistance.

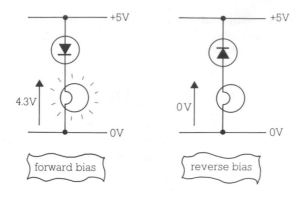

Figure 18.11 A diode only lets through it current go one way

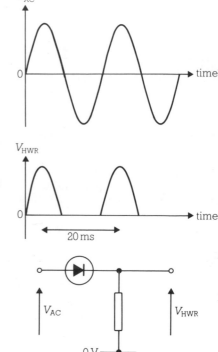

This behaviour is summarised in the diagrams of figure 18.11. When the diode is forward biased there is a 0.7 V drop across it. So there is $5 - 0.7 = 4.3$ V across the bulb on the left and a current goes through it. No current will go through a reverse biased diode so there is 0 V across the bulb on the right.

Half-wave rectifiers

A **half-wave rectifier** is a voltage divider which contains a diode and a resistor. It is shown in figure 18.12.

If V_{AC} is positive the diode will be forward biased and have a low resistance. So V_{HWR} will be roughly the same as V_{AC}.

But when V_{AC} is negative the diode will be reverse biased and have a very large resistance. So V_{HWR} will be pulled up to 0 V by the resistor.

Smoothing

The system of figure 18.12 converts an alternating voltage (V_{AC}) into one which is never negative (V_{HWR}). It feeds out positive pulses, one every 20 ms. The waveform is clearly too bumpy to be used as a supply voltage. It needs **smoothing**.

A diode in series with a capacitor can be used to convert an alternating voltage into a smoothed one. The arrangement is shown in figure 18.13. V_{DC} is now positive all of the time, even if it still wobbles up and down a bit.

Figure 18.12 A half-wave rectifier

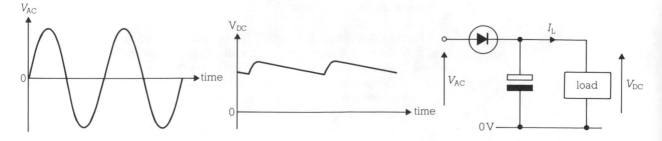

Figure 18.13 Smoothing the output of a rectifier

Each time that V_{AC} reaches its maximum voltage the diode will be forward biased. Current will go through the diode, charging up the capacitor. So V_{DC} is rapidly pulled up to within 0.7 V of V_{AC}.

As soon as V_{AC} drops below its peak value the diode becomes reverse biased. No current will go through it. The capacitor will therefore have to supply the current I_L which goes through the load. So V_{DC} will drop steadily downwards as the capacitor discharges.

Ripple

Of course, V_{AC} reaches its peak value once every 20 ms. So the capacitor gets charged up from the AC supply via the diode at 20 ms intervals. So V_{DC} rises sharply every 20 ms, with a steady fall in between. This variation in V_{DC} is known as **ripple**. Obviously, a good power supply has only a small amount of ripple.

For example, look at the power supply circuit of figure 18.14. We want it to put a steady 12 V across the load (a circuit which is run off the power supply). Have we used a large enough capacitor if the load is going to draw a current of about 100 mA from the supply?

The AC input is a 9 V rms waveform, with a peak value of $1.414 \times 9 =$ 12.7 V. So every 20 ms V_{OUT} is pulled up to $12.7 - 0.7 = 12.0$ V. If the ripple is to be small, the time constant of the capacitor/load circuit must be much larger than 20 ms. (A capacitor will be almost completely discharged after three time constants.)

The load draws 100 mA from a 12 V supply so it has a resistance of roughly 100 Ω (from Ohm's law). So the time constant RC will be about $0.1 \times 2000 = 200$ ms. As this is ten times larger than 20 ms the ripple will be fairly small.

**Increasing the current drawn from the power supply
output increases the size of the ripple.
Increasing the size of the capacitor
reduces the amount of ripple.**

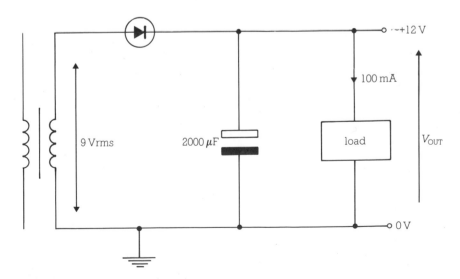

Figure 18.14 A simple power supply

Full-wave rectification

The output voltage of a good power supply should be rock steady. It should not have any ripple. In order to keep the ripple small the smoothing capacitor has to be large. But there is a limit to how large you can make the capacitor (about 10 000 μF). A better strategy is to arrange things so that the capacitor is recharged once every 10 ms instead of once every 20 ms. This involves **full-wave rectification** of the AC supply.

The diode bridge

One way of doing this is to use a **diode bridge**. This arrangement of four diodes is shown in figure 18.15. The bridge has four terminals. The waveform to be rectified (V_{AC}) is fed into the two terminals marked ~. The rectified waveform (V_{FWR}) appears at the other two terminals marked + and −. The graph of figure 18.15 shows that V_{FWR} is always positive. The lumpiness could obviously be ironed out by placing a capacitor in parallel with the load.

A simple power supply

A simple power supply which converts the 9 V rms AC output of a transformer into a smooth + 11 V is shown in figure 18.16. The load represents the electronic systems which are being run off the power supply.

The voltage across the load can be worked out as follows. The peak voltage of the transformer output is 1.414 × 9 = 12.7 V. The current which charges up the 1000 μF capacitor every 10 ms has to go through two diodes, so the capacitor is charged up to 12.7 − 1.4 = 11.3 V. Therefore the power supply will place about 11 V across the load.

The amount of ripple on the power supply's output will depend on how much current is drawn by the load. The bigger that current is, the worse the ripple will be. You can halve the amount of ripple by doubling the size of the capacitor.

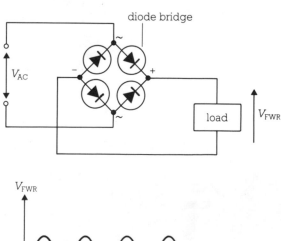

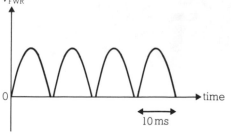

Figure 18.15 Using a diode bridge for full-wave rectification

Figure 18.16 A simple power supply

QUESTIONS

1 A mains cable contains three colour coded wires. Give the names of the wires and their colours. Which of them goes to the fuse in a plug?

2 Copy and complete the following statements.
The .... wire in a mains cable carries a dangerous ...... voltage. Both the ..... and ...... wires carry the alternating ...... Every device connected to the mains supply should have a ..... and a fuse in the ..... line. A .... bulb should be connected between the .... and ..... lines to show when the mains supply is connected. The earth wire should be connected to ....

3 A mains transformer has a power rating of 6 W. If it converts the 240 V rms mains supply to 12 V rms, what is the maximum current it can deliver safely? Draw a diagram to show how you would connect the transformer up to the mains electricity supply.

4 A student builds the simple power supply shown in figure 18.17. She wants to use it to run a circuit containing a number of logic gates.
   a) What is the voltage drop across the diode when it is forward biased?
   b) What does the peak value of V_{AC} have to be?
   c) What does the rms value of V_{AC} have to be?
   d) What would happen to the ripple on the 5 V output voltage if the capacitor were replaced with a 470 μF one?

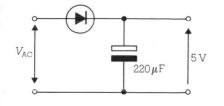

Figure 18.17 Question 4

5 Figure 18.18 is the circuit diagram of a student's project. It contains many mistakes, some of them lethal. Given that the student was trying to design a smoothed 12 V power supply, redraw the circuit diagram so that his system will operate safely and correctly.

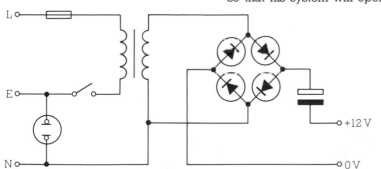

Figure 18.18 Question 5

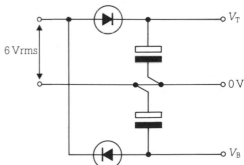

Figure 18.19 Question 6

6 This question is about the operation of split rail power supplies. The circuit is shown in figure 18.19. A 6 V rms alternating voltage is fed in from the left.
   a) What is the peak value of the alternating voltage?
   b) What is the value of V_T?
   c) What is the value of V_B?
   d) Draw the circuit diagram of a mains-driven power supply which could provide +5 V, 0 V and −5 V for running op-amps. Your circuit should include smoothing capacitors, diodes, a switch, a fuse, a neon bulb and a suitable transformer.

7 Many integrated circuits are damaged if they are connected to the supply rails round the wrong way. Figure 18.20 shows how a fuse and a diode can prevent a device being damaged in this way. Copy and complete the following statements.

> When A is at a higher voltage than B, the diode is ...... biased. So it has a .... resistance and ... current flows through it. Should A ever be at a lower voltage than B the diode would be .... biased, giving it a ... resistance. A lot of current would flow through it from the supply connected (wrongly) to A and B. That current would ... the fuse and .... the device from the power supply.

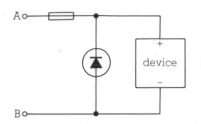

Figure 18.20 Question 7

8 This question will help you to understand how the diode bridge of figure 18.21 works. Initially T is held at + 10 V and B is held at 0 V.

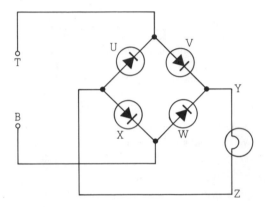

Figure 18.21 Question 8

a) Which of the two diodes U and V is forward biased?
b) What is the voltage at Y?
c) Which of the diodes X and W is forward biased?
d) What is the voltage at Z?
e) Does the current go through the bulb from Y to Z or from Z to Y?

Now repeat steps a) to e) but with T held at 0 V and B held at + 10 V. Does the current go the same way through the bulb regardless of which of the input terminals (T and B) is at the higher voltage?

19
Regulated power supplies

The simple power supply shown in figure 19.1 has two main defects. Firstly, the output voltage develops a ripple when current goes through the load. Secondly, the size of the output voltage depends on the amplitude of the alternating voltage fed into the transformer primary. Both defects mean that the system is not really suitable for providing power for integrated circuits. **Its output lacks regulation.**

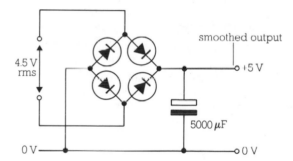

Figure 19.1 An unregulated power supply

Regulators

The block diagram of a power supply which has a regulated output is shown in figure 19.2. The first three blocks are already present in the circuit of figure 19.1. The output of the smoother (usually a large capacitor) is used to run a regulator. This feeds out a steady DC voltage of a fixed value which does not develop any ripple when current is drawn from it.

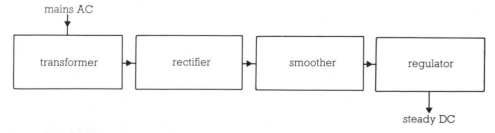

Figure 19.2 Block diagram of a regulated power supply

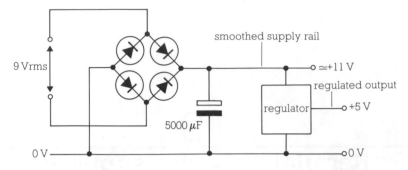

Figure 19.3 Using a regulator

A circuit diagram of a regulated supply is shown in figure 19.3. The 9 V rms AC output of a transformer is rectified by a diode bridge and smoothed by a 5000 μF capacitor to provide the top supply rail for the regulator. That smoothed supply rail will be at about +11 V, but may have a few volts ripple on it. The output of the regulator is a supply rail at a rock steady +5 V.

Zener diodes

A regulator needs to be able to generate a **reference voltage**. The most convenient way of doing this is to employ a zener diode.

The *I-V* characteristic of a typical zener diode is shown in figure 19.4.

When it is forward biased it behaves like an ordinary silicon diode. The current *I* increases dramatically when the voltage *V* exceeds +0.7 V.

When *V* is negative the diode is reverse biased. The current is zero if the voltage is less than the **breakdown voltage** of the diode. This is 2.7 V in our example. If the reverse voltage exceeds the breakdown voltage the current through the zener diode increases very rapidly.

So when current goes through a reverse biased zener diode the voltage across it will be almost the same as its breakdown voltage.

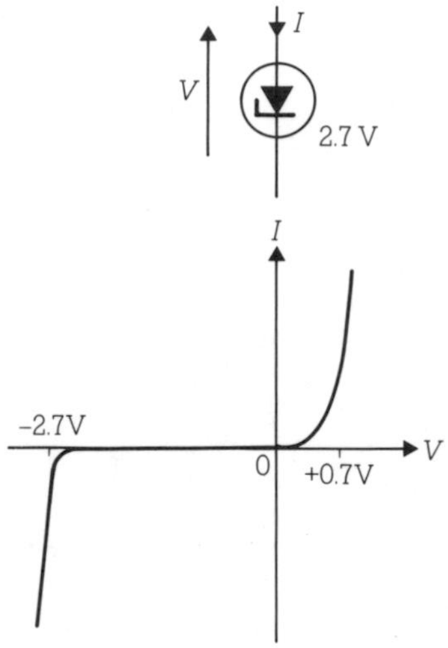

Figure 19.4 *I–V* characteristic of a zener diode

Voltage references

Figure 19.5 shows how a zener diode can be used to generate a reference voltage from a smoothed supply rail.

The 470 Ω resistor feeds current from the top supply rail into a reverse biased zener diode. Its breakdown voltage (usually printed on it) is 5.1 V, so the voltage across it (V_{REF}) will be just above 5 V. The graph shows that V_{REF} remains almost constant provided that less than about 10 mA goes through the load.

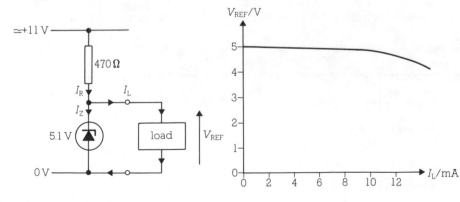

Figure 19.5 Using a zener diode to generate a reference voltage

The circuit of figure 19.5 is a simple regulator. Provided that the zener current I_Z is not zero, then V_{REF} will be about $+5$ V. How big does the load current I_L have to be before V_{REF} falls below $+5$ V?

We can use Ohm's law to calculate the current I_R which goes through the resistor.

$$R = \frac{V}{I} \qquad \begin{matrix} R = 0.47 \text{ k}\Omega \\ V = 11 - 5 = 7 \text{ V} \\ I = ? \end{matrix} \qquad \text{therefore } 0.47 = \frac{7}{I_R}$$

$$I_R = \frac{7}{0.47} = 13 \text{ mA}$$

The current through the resistor will remain a steady 13 mA when V_{REF} is at $+5$ V. Since $I_R = I_Z + I_L$, the current through the load can go up to 13 mA before the zener current becomes zero and V_{REF} starts to fall.

Power ratings

A regulator which can only supply a current of up to about 10 mA is not much use. How do you increase the output current without upsetting the regulation?

The zener diode in figure 19.6 has a **power rating** of 500 mW. We can therefore calculate the maximum safe zener current.

$$W - VI \qquad \begin{matrix} W = 500 \text{ mW} \\ V = 5.1 \text{ V} \\ I = ? \end{matrix} \qquad \text{therefore } 500 = 5.1 \times I$$

$$I = \frac{500}{5.1} \simeq 100 \text{ mA}$$

Ohm's Law can now be used to calculate the resistance which will feed this current into the zener diode from the $+11$ V supply rail.

$$R = \frac{V}{I} \qquad \begin{matrix} R = ? \\ V = 11 - 5 = 7 \text{ V} \\ I = 100 \text{ mA} \end{matrix} \qquad R = \frac{7}{100} = 0.07 \text{ k}\Omega$$

The nearest preferred value for R is 82 Ω. (68 Ω would have let too much current go through the zener diode.) So the circuit of figure 19.6 could act as a regulator capable of supplying about 100 mA of current at $+5$ V.

Regulators

The circuit diagram of a regulator which could provide a current of up to 1 A at $+5$ V is shown in figure 19.7.

The op-amp and npn transistor act as a power amplifier for the output of the zener diode voltage reference. The negative feedback applied to the op-amp ensures that V_{OUT} is the same as V_{REF}.

Note how the op-amp uses the $+11$ V and 0 V supply rails for its power supply. The op-amp will be able to provide a base current of up to about 10 mA for the transistor. If the current gain is 100, the emitter current should be able to go up to $10 \times 100 = 1000$ mA. This will generate heat at a rate of $VI = (11 - 5) \times 1000 = 7000$ mW or 7 W, so the transistor will definitely need a heat sink to keep it cool.

The whole circuit of figure 19.7 could be produced in integrated circuit form. For example, the 7805 IC regulator can convert a smoothed voltage of between $+7.5$ V and $+25$ V into a steady $+5$ V at a current of up to

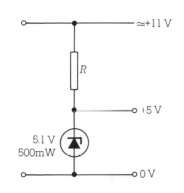

Figure 19.6 A simple regulator

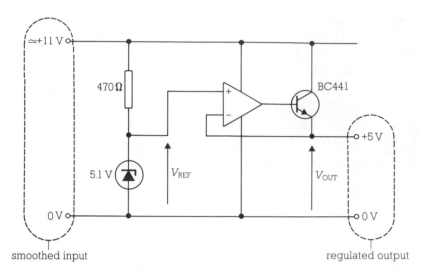

Figure 19.7 A 5 V regulator

1 A. Furthermore, it contains circuitry to switch it off if it gets too hot or the current drawn from it exceeds 1 A. But although the active components can be miniaturised, the whole system has to be kept bulky with a heat sink to keep it cool.

QUESTIONS

1 The zener diode in the circuit of figure 19.8 has a breakdown voltage of 9.1 V and a power rating of 1300 mW.
   a) Calculate the maximum safe current which can go through the diode.
   b) Work out a suitable preferred value for the resistor which would allow the maximum current through the diode.
   c) Sketch a graph to show how V_{OUT} would change with I_L. Mark both axes carefully.

2 Explain what a voltage regulator is supposed to do.

3 You have to design the circuit of a mains-driven regulated +9 V supply using the following components. Transistor, resistor, zener diode, transformer, diode bridge, capacitor, op-amp, switch, fuse and neon bulb.
   a) Draw a circuit diagram for the system.
   b) The transformer secondary feeds out a 12 V rms waveform. If you want a current of about 10 mA through the 9.1 V zener diode, choose a suitable value for the resistor.
   c) Explain why the output voltage of your circuit remains a steady +9 V when the mains voltage drops from 240 V rms to 210 V rms.
   d) Would you use a 10 000 μF capacitor or a 100 μF capacitor? Explain why.

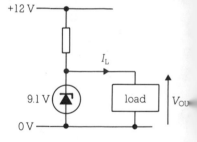

Figure 19.8 Question 1

Revision questions for Section D

1   Look at the system shown in figure D.1
    a)   Name components A, B and C. Explain what they do.
    b)   B has a voltage gain of +20. Explain what this means.
    c)   A produces a 1 kHz sine wave with an amplitude of 20 mV. If that
         signal is fed into B, what signal gets fed into C?
    d)   Suggest a practical use for the circuit of figure D.1.

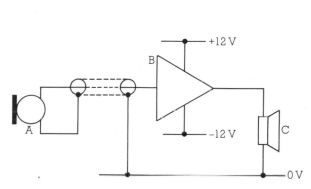

Figure D.1 Question 1

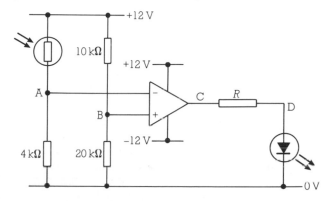

Figure D.2 Question 2

2   The circuit of figure D.2 acts as a light level detector. The LED comes
    on if the LDR is in the dark.
    a)   Copy and complete the following statements.
         The output of the op-amp can get to within about . . . . V of either
         supply rail. So when A is at a higher voltage than B, C is . . . . V. If
         A is at a lower voltage than B, C is . . . . V.
    b)   Calculate the voltage at B.
    c)   What range of voltages can A have if the LED is
         i)   on,
         ii) off?
    d)   What is the resistance of the LDR when the LED is about to go
         off?
    e)   The LED is rated at 2 V, 5 mA. Calculate a suitable preferred
         value for the resistor R.

3   This question is about the circuit of figure D.3.
    a)   What happens to the voltage at S when the switch is
         i)   pressed,
         ii) released?
    b)   Suppose that T is set to +2 V. What does the voltage at S have to
         be to make the buzzer
         i)   make a sound,
         ii) stay silent?
    c)   When the switch is released, there is a time delay before the
         buzzer stops making a sound. What happens to the length of that
         time delay if T is
         i)   lowered to +1 V,
         ii) raised to +4 V?

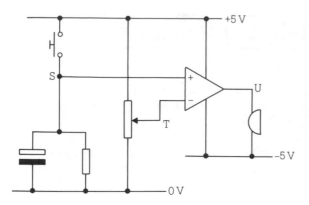

Figure D.3 Question 3

4 An op-amp which is run off split supply rails of $+12$ V and -12 V is shown in figure D.4. The two resistors make it into an amplifier.
  a) Is the amplifier an inverting or a non-inverting one?
  b) Calculate its voltage gain.
  c) Complete the table below.

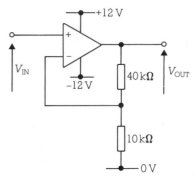

Figure D.4 Question 4

V_{IN}/V	0	$+1$	$+2$	$+3$	-1	-2
V_{OUT}/V	?	?	?	?	?	?

5 Draw circuit diagrams to show how op-amps can be made into amplifiers which have voltage gains of $+23$ and -47. Make the smaller resistor 10 kΩ each time.

6 This question is about the audio system shown in figure D.5.
  a) State the range of sound frequencies which human beings can hear.
  b) The system contains an inverting amplifier and a power amplifier. Which section is which, and what is their function?
  c) What is the gain of the inverting amplifier?
  d) A sine wave of amplitude 20 mV is fed in at IN. How big a signal appears at
     i) B,
     ii) OUT?

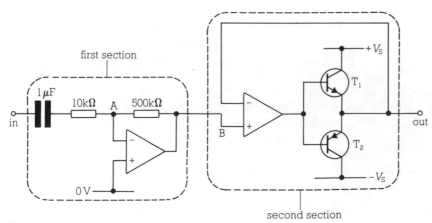

Figure D.5 Question 6

e) Suppose that B is at + 1 V. Which of the transistors is on? Which one comes on when B is at − 2 V?
f) The system of figure D.5 has a fixed gain. The addition of a potentiometer would allow its gain to be varied. Show how this could be done.
g) Is the system suitable for amplifying the output of
   i) a crystal microphone (impedance 47 kΩ),
   ii) a moving-coil microphone (impedance 1 kΩ)?

7 Draw the circuit symbol for an npn transistor. Label the base, emitter and collector. Complete the following.
To turn the transistor on the base must be .... V above the emitter. When the transistor is on, current can go from the collector to the ..... To turn the transistor off, the base must be ..... When it is off, no current can flow from the .... to the .....

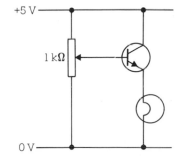

+5 V
1 kΩ
0 V

Figure D.6 Question 8

8 Figure D.6 shows a transistor controlling the current going through a light bulb. Its current gain is 100.
a) If 50 mA goes through the bulb, what is the value of
   i) the emitter current,
   ii) the base current,
   iii) the collector current?
b) If the voltage across the bulb is 2.0 V, what is the value of
   i) the emitter voltage,
   ii) the base voltage,
   iii) the collector voltage?
c) Draw a graph to show how the voltage across the bulb depends on the voltage of the wiper. Mark on your graph the regions where the transistor is on and off.

9 Information can be sent from one place to another along a pair of wires. It can be transmitted in the form of an audio signal or as a series of pulses in Morse code. What are the advantages of using the audio signal for communication?

10 Describe, with the help of some examples, the way in which the modern pop music industry relies on electronics.

11 Figure D.7 shows three rectifiers. The graph shows how the voltage fed into them (V_{IN}) varies with time. Draw graphs to show how V_{OUT} changes with time for each of the circuits.

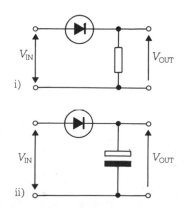
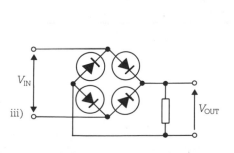
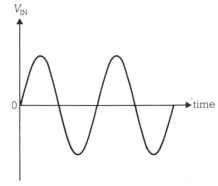

Figure D.7 Question 11

12 A power supply which uses the 240 V rms mains supply to generate a steady +9 V is shown in figure D.8.

    a) Draw a diagram to show how the live, neutral and earth lines should be connected to the terminals A, B and C. Include a switch, a neon bulb and a fuse.

    b) State the colours of the live, earth and neutral wires in a mains cable.

    c) Name the component marked S. Describe its behaviour.

    d) The point marked T has a peak voltage of +15 V and a ripple of 5 V. The mains supply has a frequency of 50 Hz. Draw a graph to show how the voltage at T changes with time.

    e) The capacitor is replaced with a 5000 μF one. What effect does this have on

        i) the peak voltage at T,

        ii) the ripple at T?

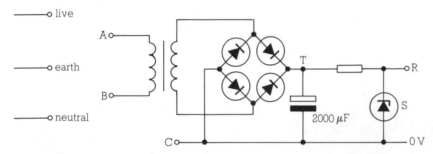

Figure D.8 Question 12

13 Draw a block diagram of a regulated power supply which converts the mains AC supply into a steady 12 V. Describe the function of each block.

14 A 9 V battery is used to make a 5 V power supply with the help of the circuit shown in figure D.9.

    a) What is the voltage drop across the resistor?

    b) How much current flows through the resistor?

    c) If the current drawn from the 5 V supply rail (I_{OUT}) is 15 mA, how much goes through the zener diode?

    d) What is the maximum current which can be drawn from the 5 V supply rail without affecting its voltage? Explain your answer.

    e) Draw a circuit diagram to show how an op-amp and a transistor could be used to increase the current available from the 5 V output. Explain how your system works.

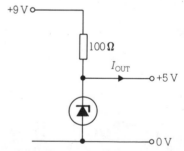

Figure D.9 Question 14

Counters, clocks
controllers and converters

An electronic control system lets
this one machine do the work of
many other machines. A piece of
metal fits in the holder on the left.
The tools on the upper and lower
blocks then cut the metal to any
shape you want. *(NCMT)*

20
Binary counters

A **counter** is a special type of memory. A memory just remembers binary words which have been written into it. Exactly what that word represents depends on how it has been coded. It could represent a number, a letter of the alphabet or anything else.

The word stored by a counter represents only one thing, namely the number of pulses which have been fed into it.

A counting system

A system which can count up to 99 **events** is shown in the block diagram of figure 20.1. The system could be used to count a large variety of events. It might count the people passing through a turnstile, the number of cars going down a road or how many packets of ice cream pass down a conveyor belt.

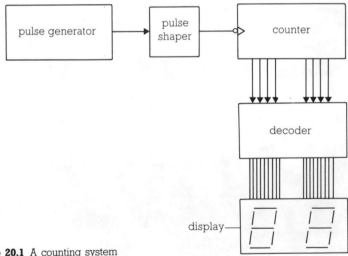

Figure 20.1 A counting system

The **pulse generator** produces a pulse each time that an event happens. There are many ways in which events can be used to make pulses. All you need is a suitable input transducer. For example, the pulse generator could be a push switch. An event would then be anything which closed the switch.

Each event, whatever it is, will make the pulse generator feed out a pulse. Counters are very fussy about the shape of the pulses which they count, so a **pulse shaper** is used to clean up each pulse before it gets to the counter.

The outputs of the counter feed out a binary word which represents how many events have happened. That number increases by one every time another pulse enters the counter.

Finally, a **decoder** reads in the counter's outputs and drives some sort of **display**.

Clean pulses

For a pulse to be counted successfully its edges must be very sharp. The pulse must rise very rapidly from 0 to 1 and drop equally rapidly from 1 to 0.

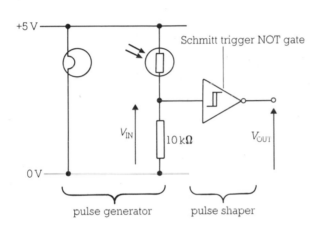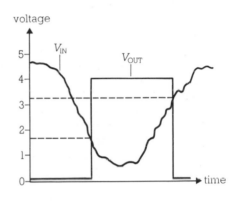

Figure 20.2 Using a Schmitt trigger NOT gate to clean up clock pulses

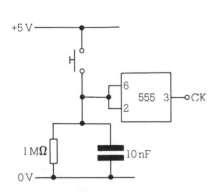

Figure 20.3 A switch circuit which generates clean pulses

The pulses which come out of most pulse generators are rarely suitable for feeding directly into a counter. For example, look at the system shown in figure 20.2. Each time that something is placed between the bulb and the LDR, V_{IN} will fall from +5 V to 0 V. When the object is removed again, V_{IN} will rise back up to +5 V. As the graph shows, V_{IN} doesn't rise or fall rapidly. It also wiggles around a lot i.e. it is noisy.

A Schmitt trigger NOT gate is used as a pulse shaper in figure 20.2 to clean up V_{IN}. When V_{IN} gets to the lower trip point, V_{OUT} shoots up rapidly from 0 to 1. Similarly, when V_{IN} gets up to the upper trip point, V_{OUT} shoots down rapidly from 1 to 0.

The sharp rising and falling edges of V_{OUT} mean that it will be correctly interpreted as a single pulse by a counter. V_{IN} might have been interpreted as several pulses or simply ignored altogether!

A useful system for obtaining clean pulses from a push switch is shown in figure 20.3. The 1 MΩ resistor and 10 nF capacitor eliminate any bounce effects so that CK goes immediately and cleanly from 1 to 0 each time that the switch is pressed. The 555 IC is, of course, a Schmitt trigger NOT gate.

Making D flip-flops count

Since a counter is a special type of electronic memory, it will come as no surprise that it can be made out of one bit memories or D flip-flops.

One bit counters

If you connect the $\overline{Q}$ output of a D flip-flop to its D input (see figure 20.4) you end up with a system with the following behaviour.

Q changes state each time a rising edge enters CK.

This is shown in the timing diagram of figure 20.4. Each time that IN rises from 0 to 1, OUT becomes whatever it wasn't before. If it was a 0 it becomes a 1. Or if it was a 1 it becomes a 0.

Feedback

It is not difficult to see why the system behaves this way. Figure 20.5 shows the two possible states for the system, with Q a 0 on the left. Since $Q \neq \overline{Q}$, $\overline{Q}$ must be 1 if Q is 0. The feedback path carries the state of $\overline{Q}$ round to D. Since a rising edge fed into the clock terminal will make Q the same as D ,that pulse will change Q from 0 to 1.

Now we have the situation shown on the right of figure 20.5. $\overline{Q}$ is 0, so the next rising edge fed into the clock terminal will make Q go back to 0.

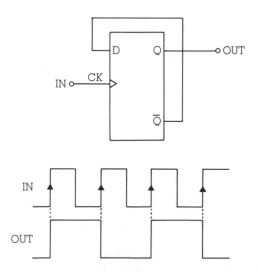

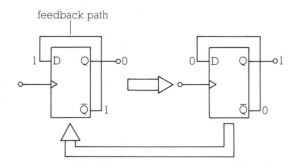

Figure 20.4 A one bit counter

Figure 20.5 The two states of a one bit counter

Two bit counters

The one bit counter of figure 20.4 is the basic unit of electronic counters. By putting enough one bit counters in series with each other you can make systems which can count as many events as you like.

For example, the two bit counter shown in figure 20.6 will count up to three pulses. Those pulses are fed into the clock input of the left hand D flip-flop. B and A are the outputs of the counter. The binary word BA represents how many pulses have been fed into CK, according to the table below.

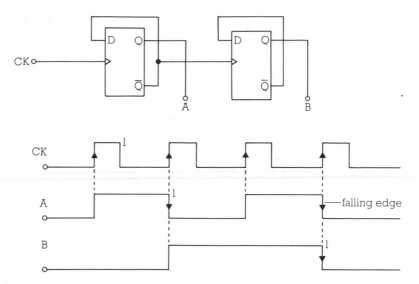

Figure 20.6 A two bit counter

Pulse number	B	A
0	0	0
1	0	1
2	1	0
3	1	1
4	0	0
5	0	1

Note how the fourth pulse returns the word BA to its original state of 00. Obviously BA can represent, in binary, a number between 0 and 3. So as pulses enter CK the system counts up from 0 to 3 before going back to 0 and starting again.

The timing diagram of figure 20.6 is another way of showing what is going on. The output of the first flip-flop (A) changes state every time that CK rises from 0 to 1. But every time that A falls from 1 to 0 a rising edge is fed into the clock input of the second flip-flop, causing B to change state.

Large counters

A two bit counter has four different output states. (BA can be 00, 01, 10 or 11.) It can count up to $2^2 - 1 = 3$ events. So a counter with N outputs (an N bit counter) will have $2^N - 1$ states.

The 4024 IC

The circuit diagram of a seven bit counter is shown in figure 20.7. It consists of seven one bit counters connected in series with each other. It can count up to $2^7 - 1 = 127$ events, so its output is a seven bit word (GFEDCBA) which represents a number between 0 and 127.

The whole system is available in a single package, the 4024 IC (figure 20.8).

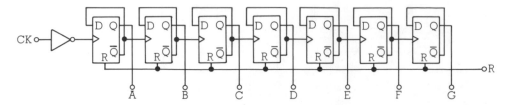

Figure 20.7 A seven bit counter

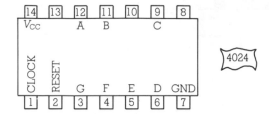

Figure 20.8 The 4024 IC

Resetting counters

The counter circuit of figure 20.7 contains a reset terminal marked R. Provided that R is a 0, the word GFEDCBA represents how many pulses have been fed into CK. But if R is a 1 then all of the flip-flops are reset and GFEDCBA is 0000000 regardless of what enters CK.

Counting falling edges

The circle next to the clock pulse input of the counter symbol shown in figure 20.9 means that it is triggered by falling edges. In other words, the system counts falling edges fed into CK; this is why the NOT gate was included in figure 20.7.

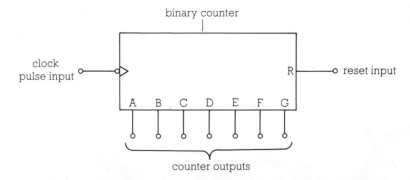

Figure 20.9 The symbol of a seven bit counter

The timing diagram of figure 20.10 shows how each stage of a useful binary counter will change state when the previous stage falls from 1 to 0. So A changes state each time CK falls, B changes state each time A falls and C changes state each time that B falls.

Each waveform changes state when the one above it falls.

This pattern ensures that the system counts up in binary as shown in the table below.

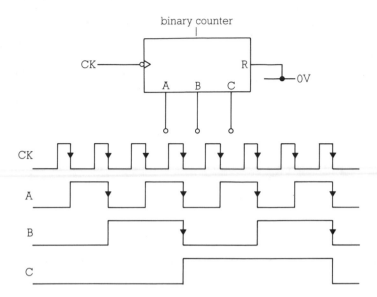

Figure 20.10 A three bit counter's timing diagram

Pulse number	C	B	A
0	0	0	0
1	0	0	1
2	0	1	0
3	0	1	1
4	1	0	0
5	1	0	1
6	1	1	0
7	1	1	1
8	0	0	0

Counters in series

Figure 20.11 shows how two three bit counters have to be connected to make a six bit counter. The output which has the slowest changing

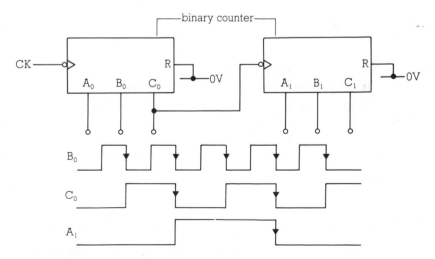

Figure 20.11 Making a six bit counter out of two three bit counters

waveform of the first counter is used to trigger the second counter. So each time that C_0 falls A_1 changes state. The six bit word $C_1B_1A_1C_0B_0A_0$ therefore represents how many pulses have been fed into CK.

Counters which are triggered by falling edges can be put in series with each other to make larger counters. A single 4024 IC will count up to $2^7 - 1 = 127$ events, so two in series would allow you to count up to $2^{14} - 1 = 16383$ events!

QUESTIONS

1   Figure 20.12 shows a one bit counter and part of its timing diagram.
  a)   What happens to the state of OUT when CK
    i)   rises from 0 to 1,
    ii)   falls from 1 to 0?
  b)   What do you have to do to IN to make OUT change state? Explain your answer.
  c)   Copy and complete the timing diagram; OUT is initially a 0.

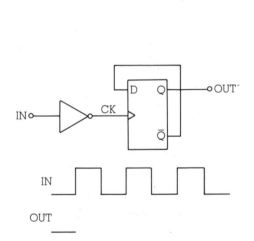

Figure 20.12 Question 1

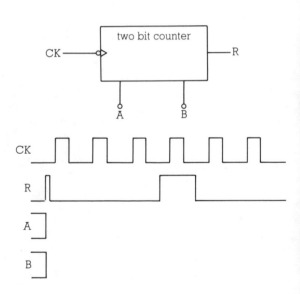

Figure 20.13 Question 3

2   How many events can a five bit counter count? How many four bit counters do you need if you want to count up to a thousand events?

3   A two bit binary counter is shown in figure 20.13. The binary word BA gives the value of the count.
  a)   Draw a circuit diagram to show how the system could be made out of a NOT gate and two D flip-flops. Label the R, CK, A and B terminals.
  b)   What is the value of BA when R is 1?
  c)   Copy and complete the timing diagram of figure 20.13.
  d)   Draw a circuit diagram to show how three two bit counters could be connected to make a six bit counter.

4 A four bit counter is reset so that its output DCBA is 0000. Pulses are then fed in until DCBA is 0101.
  a) How many pulses were fed into it?
  b) What will DCBA become after
       i) three more pulses,
       ii) ten more pulses?

5 This question is about the use of binary counters to generate low frequency waveforms from high frequency ones. A 4 kHz square wave is fed into the input of a four bit counter.
  a) Draw a timing diagram to show how the waveforms at the four outputs (D, C, B and A) are related to that of the 4 kHz square wave. (Have eight cycles of the input waveform.)
  b) Copy and complete this table.

Waveform	CK	A	B	C	D
Period/ms	$\frac{1}{4}$	?	?	?	?
Frequency/kHz	4	?	?	?	?

21
Decimal counters

An electronic counter uses a binary code to represent the number of pulses which have been fed into it. That number will have to be displayed in decimal form, as that is the easiest form for humans to understand. Since seven segment displays can display numbers between 0 and 9, we need a counter which will also count from 0 to 9.

The BCD counter

A counter which will count from 0 to 9 is shown in figure 21.1. Its four outputs (D, C, B and A) could be used to drive a seven segment display via a decoder. The AND gate ensures that every tenth pulse fed into the counter will cause it to be reset to 0.

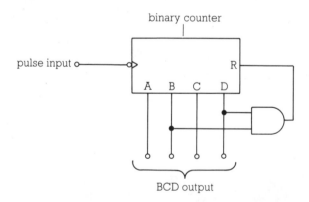

Figure 21.1 A 0 to 9 counter

Reset at ten

To understand why it behaves this way, study the truth table shown below. It shows what happens to D, C, B, A and R as pulses are progressively fed into the counter. R is generated by an AND gate connected to B and D, so it will only be 1 when both B and D are 1. So if DCBA starts off at 0000, it will get to 1001 after nine pulses. The tenth pulse makes DCBA equal to 1010. Both B and D will be 1, so R will be 1 and the counter will be immediately reset. So as soon as DCBA becomes 1010 it goes back to 0000.

The counter is reset to zero at every tenth pulse.

Pulse	D	C	B	A	R
0	0	0	0	0	0
1	0	0	0	1	0
2	0	0	1	0	0
3	0	0	1	1	0
4	0	1	0	0	0
5	0	1	0	1	0
6	0	1	1	0	0
7	0	1	1	1	0
8	1	0	0	0	0
9	1	0	0	1	0
10	1	0	1	0	1
	↓	↓	↓	↓	↓
	0	0	0	0	0
11	0	0	0	1	0
12	0	0	1	0	0

Clocks

A simple clock is shown in figure 21.2. An oscillator feeds pulses at the rate of one a second into a BCD counter of the type shown in figure 21.1. The output of the counter is decoded and used to run a seven segment display. The system counts continuously from 0 to 9.

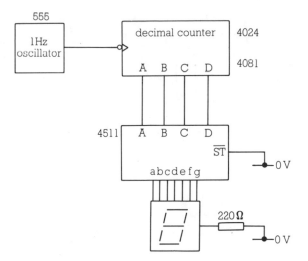

Figure 21.2 A simple seconds timer

A number of decimal counters in series will be needed to count above nine seconds. Figure 21.3 shows a system which can count up to 999 pulses. Each time that one of the counters is reset it feeds a falling edge into the next counter. So for every hundred pulses which enter the left hand counter, ten pulses are fed into the centre counter and one pulse is fed into the right hand one.

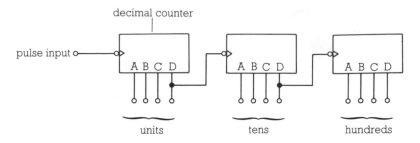

Figure 21.3 Putting decimal counters in series

By stacking decimal counters in series, you can build as large a counter as you please. Each stage can display its count via a separate decoder and seven segment display.

Hours and minutes

Figure 21.4 shows how a clock which indicates hours and minutes can be assembled.

Four counter stages are required. The oscillator feeds out one pulse per minute, so the first two counters on the left record the time in minutes. After sixty pulses have entered them they feed a falling edge into the third counter. So the two counters on the right record the time in hours. After twelve hours all of the counters are reset to 0 and the cycle starts again.

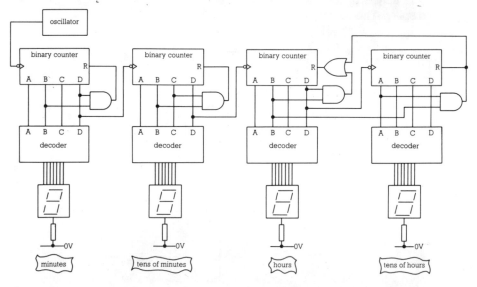

Figure 21.4 A clock circuit

Clock oscillators

A clock which is built out of counters will only keep good time if its oscillator is accurate. Unless the oscillator of figure 21.4 feeds out a single falling edge at **exactly** one minute intervals the clock will run slow or fast. Oscillators which rely on capacitors and resistors to fix their periods are not precise enough for clocks. This is because the properties of these components change as they age and as they heat up and cool down. So a clock which used a 555 IC as its oscillator would need frequent adjustment if it was to show the correct time.

There are two solutions to the problem. Use the mains electricity alternating waveform or a crystal oscillator.

Mains reference

In this country the mains electricity supply has a well regulated frequency of 50 Hz. So a transformer and Schmitt trigger can be used to generate a 50 Hz square wave suitable for counting by a clock.

For example, the circuit of figure 21.5 feeds out pulses at one second intervals. The output of the transformer is rectified by a diode and then squared with a Schmitt trigger. A pair of counters feed out one pulse for every fifty pulses fed into them from the Schmitt trigger.

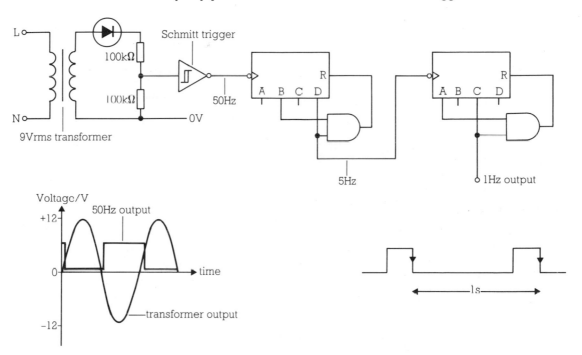

Figure 21.5 Generating a 1 Hz signal from the mains 50 Hz waveform

Crystal oscillators

Using the mains frequency to generate the pulses for a clock is only a good idea if that clock is to be run off mains electricity. Another source of precisely timed pulses must be used for portable clocks, such as watches.

Quartz crystals are widely used to generate signals of precise frequency. When it is made part of an oscillator, the crystal can force the circuit to oscillate at its own natural vibration frequency. The value of that frequency depends on the shape and size of the crystal.

Clocks and watches tend to use 32 768 Hz crystals because it is 2^{15} Hz. This is convenient because a fifteen bit binary counter can reduce the 32 768 Hz signal to a 1 Hz one.

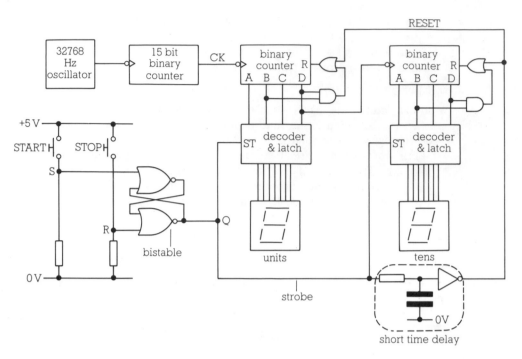

Figure 21.6 A stopwatch

A stopwatch

An example of a crystal oscillator in action is shown in figure 21.6. The stopwatch is controlled by two switches. It will start counting when the START switch is momentarily pressed. It will stop when the STOP switch is pressed. The time interval between these two events is shown, in seconds, on the two seven segment LEDs.

How it works

A fifteen bit binary counter reduces the 32 768 Hz signal from the oscillator to 1 Hz. So CK goes from 1 to 0 once every second.

The START and STOP switches control a bistable. The instant that the START switch is pressed, Q will go to 1 and stay there. A short while later the RESET line will go to 0 and the counters will commence counting the pulses at CK. As the STROBE line is a 1 the count will be displayed on the LEDs.

As soon as the STOP switch is pressed the STROBE line goes low. So the number shown on the LEDs is frozen. Soon the RESET line goes high, so that the counters are reset to 0, ignoring the pulses from CK.

The whole system can time from 0 to 99 s. It should be obvious how it could be extended to time from 0 to 999 999 s.

QUESTIONS

1 This question is about the circuit of figure 21.7.
   a) What is the value of the word DCBA when R is 1?
   b) Copy and complete this table. It shows what happens as successive pulses are fed into CK.

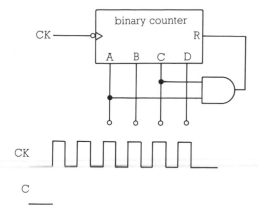

Figure 21.7 Question 1

Pulse number	D	C	B	A	R
0	0	0	0	0	?
1	?	?	?	?	?
2	?	?	?	?	?
3	?	?	?	?	?
4	?	?	?	?	?
5	?	?	?	?	?
6	?	?	?	?	?

   c) Copy and complete the timing diagram of figure 21.7.

2 You are provided with a four bit binary counter (like the one shown in figure 21.7) and a number of AND gates. Draw circuit diagrams to show how you would connect them to make systems which would count, in binary, from 0 to
     i) 10,
    ii) 3,
   iii) 8,
   iv) 6,
    v) 14.

3 You are provided with an oscillator which produces one falling edge (or pulse) per second. You need to design a counter system which will take in that signal and feed out one pulse per minute. Use two four bit binary counters and some AND gates. Draw a circuit diagram of your system. Explain how it works.

4 The mains electricity supply in the USA runs at a frequency of 60 Hz. Draw the block diagram of a system which could use that frequency to generate one pulse per second, using the following blocks; 0–5 counter, 0–9 counter, half-wave rectifier, Schmitt trigger, transformer.

5 Figure 21.8 is the circuit diagram of a counting system. Pulses to be counted are fed in at CK. The binary word CBA represents, in binary, the number of pulses which have been counted.
   a) What does the word CBA have to be to make T a 1?
   b) If T is a 1, what does CBA become?
   c) Copy and complete this table.

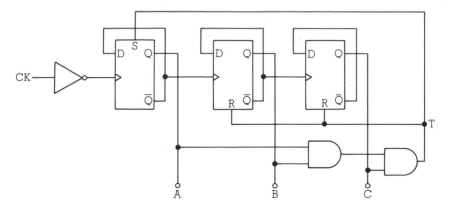

Figure 21.8 Question 5

Pulse number	C	B	A
0	1	0	0
1	?	?	?
2	?	?	?
3	?	?	?
4	?	?	?
5	?	?	?
6	?	?	?

6 Take a look at the circuit of figure 21.9. The decoder contains a four bit latch. So when ST is 1 the LEDs show the number of pulses which have entered the counter. When ST is 0 the number shown by the LEDs is frozen.

a) What do D, C, B and A have to be to make R a 1?
b) Suppose that DCBA is 0000. What will that word be after ten pulses have entered the counter? (Use a table to work it out.)
c) What do you see on the seven segment LED when
    i) P is pressed,
    ii) P is released?

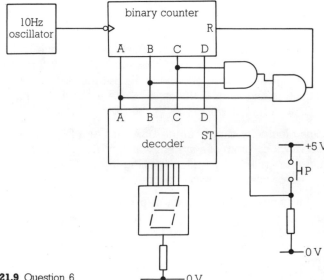

Figure 21.9 Question 6

d) The circuit of 21.9 is a random number generator. Each time that P is pressed and released a random number between 0 and 6 appears on the display and stays there. Draw a circuit diagram to show how you would adapt the circuit so that it generated random numbers between 0 and 99.

7 Although an electronic clock requires a large number of binary counters and decoders, it is possible to build one on a small piece of silicon in the form of an integrated circuit. In fact, the electronics of a digital watch can be so small that the bulkiest component of the watch is the battery!

a) Integrated circuits can be made largely automatically, but mechanical watches need to be assembled by human beings. Explain why this allows digital watches to be much cheaper than mechanical ones.

b) What other advantages do you think digital watches have over mechanical ones?

c) Mechanical watches are normally disc-shaped. What shape, if any, do digital watches have to be?

22
Sequence generators

Consider a two bit binary counter (there is one in figure 22.1). The binary word which represents the count, BA, can have one of four states (00, 01, 10 and 11). Each pulse fed into the counter changes the state of BA. So the outputs of the counter follow a fixed **sequence**. A goes 010101010101 and B goes 001100110011.

This chapter is going to show you how to build systems which follow different sequences. Such systems are very useful for controlling machinery such as washing machines, central heating systems and traffic lights.

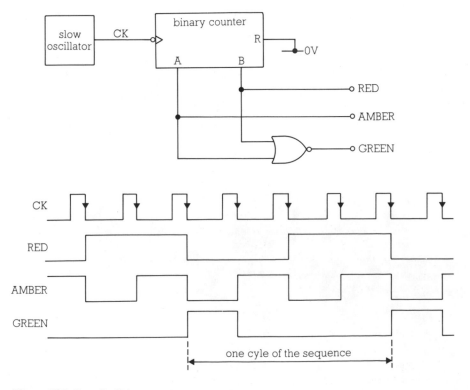

Figure 22.1 A traffic lights sequence generator

A traffic lights sequencer

Suppose that you have a red bulb, a green bulb and an amber bulb. How would you make them light up in the sequence used by traffic lights?

The problem

The sequence to be followed is this. First the green bulb comes on. Then the amber one on its own. Then the red one on its own. Then both red and amber together. Finally, return to green only and start the sequence again.

The sequence has four states. We shall number them from 0 to 3, as shown in the table in the margin.

State number	Lit bulbs
0	Green
1	Amber
2	Red
3	Red and Amber

The answer

To generate the four states, a two bit binary counter will be needed. If it is fed pulses from an oscillator it will count continuously from 0 to 3.

The truth table below shows how the outputs of the sequencer (RED, AMBER and GREEN) are to be related to the outputs of the binary counter (B and A). We have assumed that when an output is 1, its bulb will be on.

B	A	RED	AMBER	GREEN
0	0	0	0	1
0	1	0	1	0
1	0	1	0	0
1	1	1	1	0

Now consider each of the outputs in turn. If you compare the RED column with the A and B columns, it should be obvious that RED and B are the same. Similarly, AMBER and A are the same. GREEN is 1 only when both A and B are 0. So GREEN is A NOR B.

The system

The sequencer is shown in figure 22.1. The A and B outputs of the binary counter provide the RED and AMBER signals directly. A NOR gate combines A and B to generate GREEN. (The three output signals will need some buffering if they are to drive useful output transducers. You could use relays, for example, to allow those signals to control some large bulbs like those used in real traffic lights.)

The timing diagram shows that every time CK falls from 1 to 0 the system moves on to the next state in the sequence. The same amount of time is spent in each state. The amount of time it takes for the system to go through the whole sequence clearly depends on the frequency of the oscillator which produces CK.

Complex sequences

The traffic lights system was relatively simple. It only needed three outputs and there were only four steps in the sequence. Useful sequencers, such as washing machine controllers, are more complex, with many more outputs and steps in the sequence. The design of complex sequencers can, however, be quite straightforward if you use a **ROM** (read-only-memory).

ROM-based controllers

A ROM stores a number of binary words, each at a different address. Those words can be fed out of the ROM one after the other in a sequence if a binary counter is used to generate the ROM's addresses.

Washing machine controller

Figure 22.2 shows a four bit binary counter generating the addresses for a 16×6 bit ROM. Each of the ROM outputs control (via a suitable buffer) part of a washing machine. For example, D_5 controls the tap which lets water into the machine.

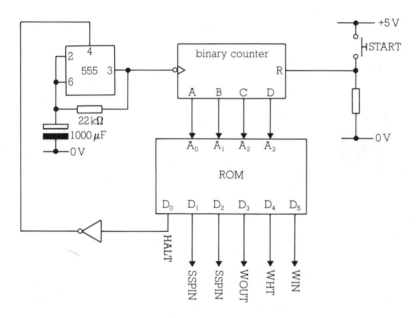

Figure 22.2 Using a ROM to generate the control sequence for a washing machine

The ROM has to control a number of devices in the washing machine. There is the water inlet valve, the water pump, the water heater, the drum motor (fast and slow). The ROM also has to stop the whole system at the end of the sequence. The function of each of the ROM's outputs is shown in the table below.

Output	Name	Function
D_0	HALT	Turn off the oscillator
D_1	FSPIN	Make the drum rotate quickly
D_2	SSPIN	Make the drum rotate slowly
D_3	WOUT	Pump water out of the machine
D_4	WHT	Heat the water
D_5	WIN	Let water into the machine

We can assume that when a ROM output goes high it activates the device which it is controlling.

The sequence

A useful washing machine cycle might go through the twelve steps shown.

Step number	Operation
0	Open the water inlet valve
1	Switch on the water heater
2	Carry on heating the water
3	Turn the drum round slowly
4	Do nothing at all
5	Turn the drum round slowly
6	Do nothing at all
7	Pump out the water
8	Let water in and turn slowly
9	Pump out and spin fast
10	Let water in and turn slowly
11	Pump out and spin fast
12	Stop

The sequence can be started again by pressing the START switch. This resets the counter, allowing the oscillator to start producing pulses again.

The program

The table below shows which six bit word needs to be stored at each location within the ROM if the sequence above is to be followed. Convince yourself that it is correct by referring to figure 22.2 and the table of operations above.

Address $A_3A_2A_1A_0$	Data $D_5D_4D_3D_2D_1D_0$
0000	100000
0001	010000
0010	010000
0011	000100
0100	000000
0101	000100
0110	000000
0111	001000
1000	100100
1001	001010
1010	100100
1011	001010
1100	000001
1101	000001
1110	000001
1111	000001

An electronic domestic hot water central heating controller. The buttons can be used to program it to control the heater and the flow of hot water round the house at various times. *(Potterton)*

This technique could be used to generate longer sequences quite easily. Furthermore, the sequence could be easily and quickly changed by using a different ROM. Indeed, different parts of a larger ROM could be selected by a set of switches before the sequence was started. So a particular sequencer could be capable of running through a number of different sequences.

Using controllers

Sequence generators are widely used to make systems perform repetitive tasks. For example, they can make a lathe perform a sequence of operations on pieces of metal over and over again. Such a lathe can work 24 hours a day without getting tired or needing a coffee break, producing identical articles automatically. By replacing, or reprogramming the ROM in the sequencer, the same lathe can be made to produce different articles. So this sort of industrial robot can be better than a human operator. Although it may be expensive to install, its running costs will be low (electronics is very robust and long-lived), it doesn't need holidays or go sick and can be re-trained instantly.

The cheapness, reliability and flexibility of modern electronic systems means that the replacement of human operators in factories by robots is an economic necessity. Articles can be produced more reliably and more cheaply by robots. So companies which invest in automatic machinery will be able to undercut their competitors in price and drive them out of the market place.

The introduction of automatic machinery into factories inevitably causes humans to be made redundant. The jobs which are lost, however, are ones which make humans the slave of a machine, doing boring and repetitive tasks. Electronics may eventually make such soul-destroying jobs a thing of the past.

QUESTIONS

1   Look at the circuit of figure 22.3.
    a)   Write out a truth table to show how the states of X, Y and Z depend on the state of the counter outputs B and A.
    b)   Describe what the LEDs do during one cycle of the sequence.

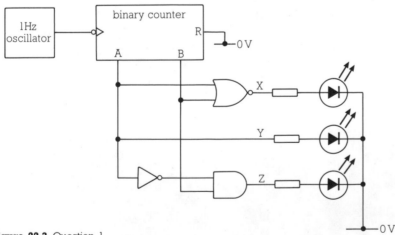

Figure 22.3 Question 1

2   A circuit has four LEDs as its output. They are labelled Q, R, S and T. They are lit for one second at a time in the following sequence; Q, R, S, T, Q, R, S, T and so on. Work out a suitable circuit for the system.

3   Explain why electronic control of machinery in factories may be better than human control. Describe some of the consequences of this.

23
Microprocessor systems

All of the electronic systems that you have met so far have been designed to perform one specific function. Their behaviour has been fixed by the way in which their components were wired together.

Hard-wired systems

For example, the logic system shown in figure 23.1 has three inputs and one output. It has been wired up so that the buzzer makes a noise whenever more than one switch is pressed. If you wanted the same components to do something else, such as make a noise when any of the switches are pressed (figure 23.2), you would have to take the circuit apart and reassemble it. The logic system is **hard-wired**. Its function cannot be altered unless its structure is altered.

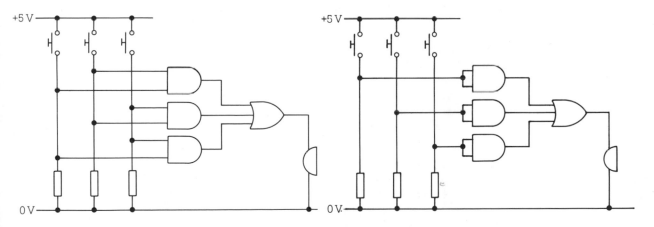

Figure 23.1 A hard-wired system   **Figure 23.2** Another hard-wired system

Programmable systems

The last chapter introduced you to a system which was not entirely hard-wired. It was the ROM-based sequence generator. The sequence that it ran through depended on the contents of its ROM. So the system can have its behaviour altered by replacing its ROM with a different one. The system is **programmable**. By replacing a single IC it can be made to behave differently.

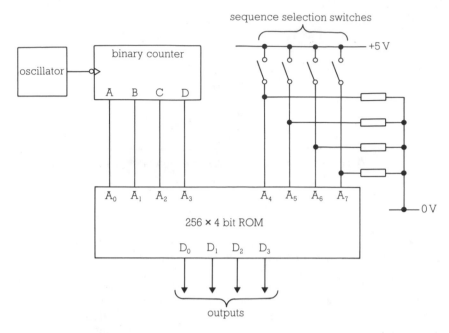

Figure 23.3 A ROM-based sequence generator

It is still a sequence generator, of course, but you can easily change the sequence by changing the ROM. You can even change the sequence by selecting a different part of the same ROM, as shown in figure 23.3.

Stored programs

The sequences stored in the ROM of figure 23.3 are somewhat like a number of shopping lists which are written down on the pages of a book. Each list shows what items you have to buy, and in what order. So the list on page 1 might be what you need for tonight's dinner, page 2 might list what you need for an electronics project, and so on.

Each shopping list is a **program**. It is a sequence of instructions to be obeyed. For example, the list on page 1 of the book might look like this.

> 4 kilos of potatoes
> 1 large meat pie
> 6 carrots
> 2 onions
> 2 kilos of apples

The program is written in a **language** which you can understand, called English. It tells you to start off by buying the potatoes, then the meat pie and so on, ending up with the apples. **It would make you perform a number of actions in a fixed sequence**.

A more realistic shopping list for tonight's dinner might contain some alternatives, in case some items were not available. It might look like this.

> 4 kilos of potatoes
> 1 large meat pie or 2 small ones
> 6 carrots
> 2 onions
> 2 kilos of apples or oranges

This time the program contains **branches**. In two places, decisions have to be made by the person carrying out the sequence. The simple sequence generator of the type shown in figure 23.3 cannot make such decisions. For an electronic system to be able to execute a program with branches, it must contain a **microprocessor**.

Microprocessors

The block diagram of a small programmable system which uses a **microprocessor** is shown in figure 23.4. Each block can be a separate IC. The behaviour of the whole system is fixed by the the binary words held in the memory. That series of words is the **program**.

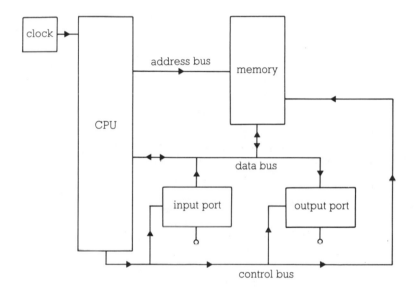

Figure 23.4 Block diagram of a simple microprocessor system

CPU operation

The microprocessor is sometimes called the **c**entral **p**rocessing **u**nit (or **CPU**) because it controls the whole system. The CPU is connected to all the other parts of the system by a number of lines or **buses**. Each bus consists of a number of parallel wires which can transport binary words from one part of the system to the other.

The CPU controls the flow of words to and from the **input** and **output ports**, dictating how the whole system is going to behave. In its turn, the CPU is told what to do by the program stored in the memory. So the CPU spends half of its time talking to the memory via the **address** and **data** buses to find out what it must do next. It spends the rest of the time moving binary words along the data bus from one block to another.

A machine cycle

Although a CPU does what it is told to by the memory, it has to follow a sequence to fetch and execute its instructions. That sequence is a **machine cycle**. It contains four steps.

1) The CPU contains a counter, known as the **program counter**, which it uses to generate the word placed on the address bus. So each time that it needs a new instruction, the CPU starts off by fetching a word from the program counter and putting it on the **address bus**.

2) The CPU then activates the memory by sending it signals along the **control bus**. This forces the memory to place one of its stored words on the **data bus**.

3) That word is then read off the data bus and placed in the **instruction register** of the CPU. A pulse is sent to the program counter so that its count increases by 1.

4) What happens next depends on the word in the instruction register. That word is an instruction, written in a language called **machine code**. It may make the CPU place a particular word in the **accumulator**, a place for storing words within the CPU itself. Or it may make the CPU feed the word in the accumulator out of the output port. Or it may make the CPU read in a word from the input port and store it in the accumulator. It may even force the CPU to change the word stored in its program counter.

   Then the CPU returns to stage 1) and sets about fetching its next instruction from the memory.

The clock

The speed with which the CPU goes through its machine cycles is fixed by the **clock**. This is a square wave oscillator which feeds timing pulses into the CPU. A popular CPU, the Z80, uses a clock frequency of 4 MHz. It can get through one machine cycle in about 2.5 µs!

A simple system

Figure 23.5 shows the **hardware** of a simple microprocessor system. It has four lines in the data bus, so its CPU handles four bit words. The whole system has four inputs (I_3, I_2, I_1 and I_0) and four outputs (O_3, O_2, O_1 and O_0). It can be programmed to behave like **any** logic system which has four inputs and four outputs. The program is the **software** of the system.

The behaviour of the hardware is determined by the software.

A simple program

For example, suppose that the ROM contained the program shown below. In practice each instruction would be represented by a four bit word.

To start the system going, the RESET switch must be released. This places 0000 in the program counter (PC). After each instruction has been fetched from the ROM and obeyed, the program counter is increased by 1. So the CPU obeys the instructions in the sequence shown.

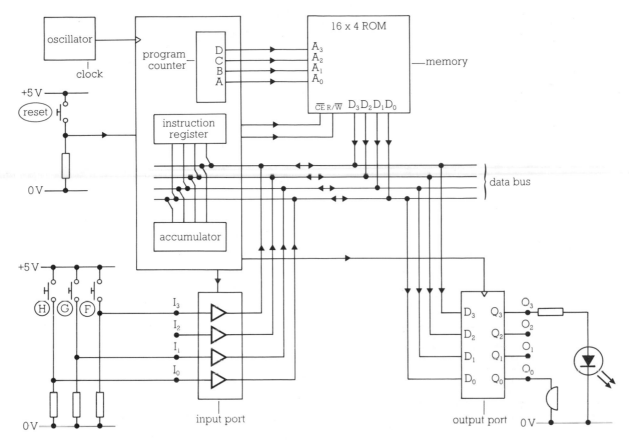

Figure 23.5 A simple microprocessor system

Address	Instruction
0000	IP → A
0001	If A ≠ 1000 then PC − 0001 → PC
0010	0001 → A
0011	A → OP
0100	IP → A
0101	If A ≠ 0011 then PC − 0001 → PC
0110	1000 → A
0111	A → OP
1000	0000 → PC

The first instruction makes the CPU place the word being fed into the input port (IP) into the accumulator (A).

The second instruction makes the CPU subtract 1 from the PC if A does not hold the word 1000. This means that this instruction is obeyed time after time until A does hold 1000.

As soon as switch F is pressed the CPU moves on to the third instruction. This makes it put the word 0001 into A.

The fourth instruction makes the CPU send the word in A to the output port (OP) where it will be frozen. This makes the buzzer go on and the LED go off.

The next two instructions make the CPU chase round in circles until both switches G and H are pressed. Then the CPU turns off the buzzer and makes the LED come on.

The final instruction loads 0000 into the PC, so that the next instruction fetched is the one at address 0000. So the CPU obeys the same set of instructions over and over again.

Larger systems

Larger microprocessor systems tend to need complex **peripherals** connected to their input and output ports. This is shown in figure 23.6, the block diagram of a small microcomputer.

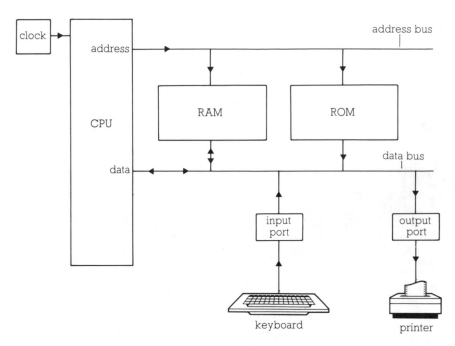

Figure 23.6 A simple microcomputer

A keyboard can be used to feed binary words into the system via the input port. The output port feeds binary words out to a printer. The ROM contains the **monitor**, a program which allows the CPU to read in the characters from the keyboard and store them in some RAM. This allows a second program to be fed into the system via the keyboard. When the appropriate command (RUN) is typed into the keyboard, the CPU uses the program in RAM rather than the program in ROM. That program can, of course feed binary words to the printer via the ouput port, getting it to print characters.

Flexible and cheap.

Microprocessors were originally developed for the production of hand-held calculators. Early microprocessor ICs were very expensive because they had cost a lot to design, and a lot of expensive research had to be done finding out how best to make them.

As we have seen, microprocessor systems are very flexible. The same **hardware** can be made to perform a wide variety of functions by giving it different **software**. So the same microprocessor and memory ICs were used in many different devices. This meant that large numbers of them were made. Once the initial development costs had been paid, this meant that the price dropped dramatically. Since ICs are made semi-automatically and use virtually no raw materials they can be produced very cheaply in large quantities.

While microprocessors were expensive their use was limited to computers and calculators. Once they became cheap, they were used in lots of other things as well. Microprocessors are now used to control washing machines, motor cars, central heating systems, burglar alarms, factory machinery, traffic lights and so on. They are being increasingly used for controlling things because they are small, cheap and robust. Above all, because the same hardware can be used in many different ways by giving it different software.

QUESTIONS

1 Draw a block diagram of a simple microprocessor system., using the following blocks:

CPU, ROM, input port, output port, clock.

Explain the function of each block of the system.

2 Explain what a bus is. Describe the function of the address, data and control buses in a microprocessor system.

3 Explain the difference between hardware and software.

4 Many microcomputers contain a Z80 CPU. So do several brands of washing machine. How can two such diverse things use the same integrated circuit?

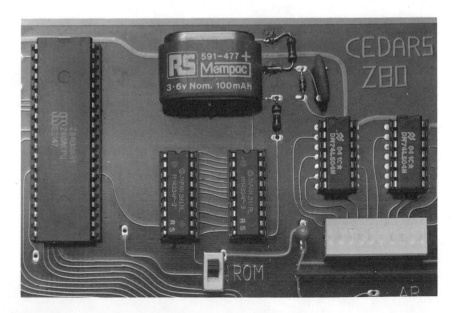

A small microprocessor system. The large IC is the CPU. The other ICs act as the memory and input/output parts.

24

Digital-to-analogue converters

A **digital-to-analogue converter** (or **DAC**) is a device which converts a binary word into a voltage. For example, the four bit DAC shown in figure 24.1 has four digital inputs (D, C, B and A) and an analogue output. The formula and table below show how the output voltage, V_{OUT}, is related to the binary word DCBA.

$$V_{OUT} = 0.1 \times (8D + 4C + 2B + A)$$

Binary word DCBA	Output voltage/V
0000	+0.0
0001	+0.1
0010	+0.2
0011	+0.3
••••	••••
1101	+1.3
1110	+1.4
1111	+1.5

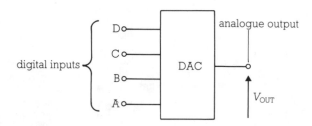

Figure 24.1 A digital-to-analogue converter

Using DACs

DACs allow complex digital systems based on microprocessors to produce analogue signals. In particular, they allow them to produce sounds.

The block diagram of a sound synthesis system is shown in figure 24.2. The microprocessor system has the DAC of figure 24.1 connected to its output port. A power amplifier allows the waveform produced at the output of the DAC to be fed into a speaker to make sound waves.

Each time that the CPU feeds the appropriate sequence of binary words in rapid succession into the DAC, a particular sound will be made by the

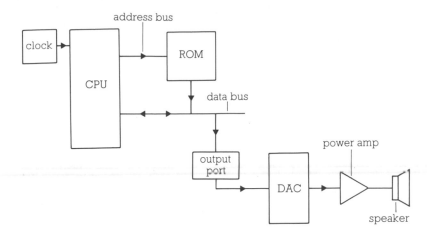

Figure 24.2 Using a DAC to generate sound

speaker. As well as holding the program, the ROM stores the sequence of binary words needed to make the sound.

For example, if the sequence of bytes shown in the table below is repeatedly fed into the four bit DAC, a sine wave is fed into the speaker. As figure 24.3 shows, it will be made up of steps, each step lasting for 0.1 ms. Those steps can be smoothed out (with the help of a capacitor) to give a reasonable approximation to a 62.5 Hz sine wave.

$$0111 \rightarrow 1010 \rightarrow 1101 \rightarrow 1110 \rightarrow 1111 \rightarrow 1110 \rightarrow 1100 \rightarrow 1001$$
$$\uparrow 0101 \leftarrow 0010 \leftarrow 0001 \leftarrow 0000 \leftarrow 0001 \leftarrow 0011 \leftarrow 0110 \downarrow$$

More bits

The steps in the sine wave of figure 24.3 are 0.1 V or 100 mV apart. This is because a four bit DAC only feeds out $2^4 = 16$ different voltages. An eight bit DAC which covered the same voltage range (0.0 to +1.5 V) would produce steps which were only $1.5 \div 256 \approx 0.006V = 6$ mV apart. This would clearly give a more precisely defined output waveform, especially if more than sixteen bytes were used to define one cycle of the waveform.

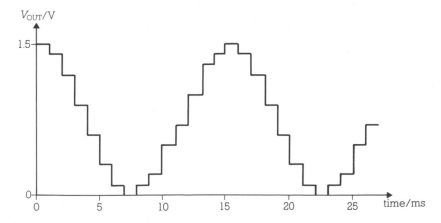

Figure 24.3 Making a sine wave with a DAC

Making speech

A microprocessor system connected to a DAC and a power amplifier can make any sound you like. All you have to do is store the appropriate sequence of bytes in ROM and get the CPU to feed them in rapid succession to the DAC. In particular, you can use the system to make the sounds of human speech. An 8192×8 bit ROM can, for example, store enough data to generate all the words necessary for a speaking clock and calendar.

The main expense of such systems is the large amount of memory they need. As memory ICs become cheaper, electronic systems which can talk to their users will become commonplace. While they are relatively expensive, speech synthesis systems will only be used in applications where expense is not important. These tend to be military and safety applications.

Where speech is useful

Talking electronic systems are useful because they can give information to their user without requiring him to move his eyes. This can save vital seconds in an emergency.

For example, a talking radar system could inform a fighter pilot about other aircraft and missiles. A talking radioactivity meter could help operators of a nuclear power station in an emergency. A talking car could tell its driver about any dangerous conditions (low oil pressure, high temperature, low air pressure and so on).

DAC construction

A DAC has to be able to combine a number of signals, each of which can be 0 V or $+5$ V, to produce its analogue output. Those input signals are combined with a **summing amplifier**.

Summing amplifiers

A simple summing amplifier is shown in figure 24.4. Two signals (V_1 and V_2) are fed into it. The output signal (V_{OUT}) is given by this formula.

$$V_{OUT} = - \left(V_1 \times \frac{R_F}{R_1} + V_2 \times \frac{R_F}{R_2} \right)$$

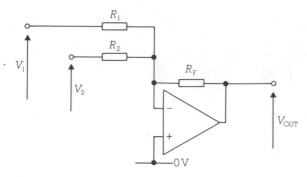

Figure 24.4 A summing amplifier

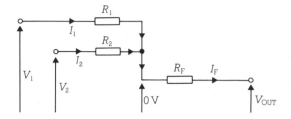

Figure 24.5 Current going through the summing amplifier

It is not difficult to see where this formula comes from. R_F provides negative feedback for the op-amp. So its inverting input is a virtual earth and sits at 0 V. Ohm's law can be used to work out how much current goes through R_1 and R_2 (see figure 24.5).

$$R_1 = \frac{V_1}{I_1} \qquad \text{therefore } I_1 = \frac{V_1}{R_1}$$

$$\text{similarly, } I_2 = \frac{V_2}{R_2}$$

The two currents I_1 and I_2 cannot go into the inverting input of the op-amp. So they have to go into the feedback resistor R_F. By applying Ohm's law to that resistor, V_{OUT} can be calculated.

$$R_F = \frac{0 - V_{OUT}}{I_F} \qquad \text{therefore } V_{OUT} = - R_F \times I_F$$

$$I_F = I_1 + I_2 \qquad \text{therefore } V_{OUT} = - R_F \times (I_1 + I_2)$$

$$\text{therefore } V_{OUT} = -R_F \times \left(\frac{V_1}{R_1} + \frac{V_2}{R_2} \right)$$

If you shuffle the symbols around in the last expression you get the summing amplifier formula quoted above.

Audio mixers

Summing amplifiers can add two or more audio signals together. Circuits which can do this are sometimes called **audio mixers**. One is shown in figure 24.6. Two signals S_1 and S_2 are AC coupled into the system. The output will be the sum of the two signals, attenuated by a factor of two.

$$S_{OUT} = - \tfrac{1}{2}(S_1 + S_2)$$

Audio mixers are widely used in recording studios to mix two or more sounds together.

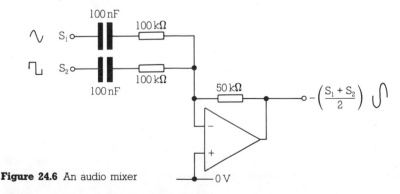

Figure 24.6 An audio mixer

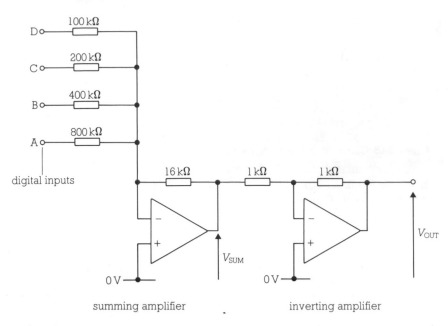

Figure 24.7 Making a DAC out of op-amps

Adding digital signals

Figure 24.7 shows a system which can add four digital signals together. The signals which are fed into D, C, B and A are supposed to be either +5 V or 0 V. (1 or 0.) The output of the first op-amp is given by the following formula.

$$V_{SUM} = - \left(5D \times \frac{16}{100} + 5C \times \frac{16}{200} + 5B \times \frac{16}{400} + 5A \times \frac{16}{800} \right)$$

where D, C, B and A are 1 or 0. (The formula is that for a normal summing amplifier with four inputs.) If we shuffle the numbers around a bit, we get the following expression for V_{SUM}.

$$V_{SUM} = -0.1 \times (8D + 4C + 2B + A)$$

Obviously V_{SUM} is proportional to the value of the binary number represented by the word DCBA. The addition of an inverting amplifier to get rid of the minus sign makes a system with all the characteristics of the DAC of figure 24.1.

$$V_{OUT} = -1 \times V_{SUM} = 0.1 \times (8D + 4C + 2B + A)$$

Integrated circuit DACs

The circuit of figure 24.7 relies on the four input resistors (100 kΩ, 200 kΩ, 400 kΩ and 800 kΩ) being in the exact ratio 1:2:4:8. This is difficult, but not impossible to achieve with discrete resistors. However, a four bit DAC is only of limited use. An eight bit DAC would need eight resistors in the exact ratio 1:2:4:8:16:32:64:128. This is virtually impossible with discrete resistors, but can be achieved fairly easily if the whole system is built as an integrated circuit. The precision with which they have to be built makes DAC ICs about twenty times more expensive than logic gate ICs.

QUESTIONS

1 What is a four bit DAC? Describe what it is supposed to do.

2 What is the output voltage of the summing amplifier shown in figure 24.8 if
   a) $V_1 = +1.0$ V and $V_2 = +1.0$ V,
   b) $V_1 = +2.0$ V and $V_2 = -2.0$ V,
   c) $V_1 = +0.5$ V and $V_2 = -1.5$ V?

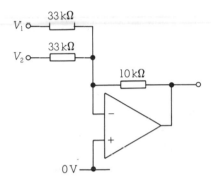

Figure 24.8 Question 2

3 A four bit DAC made out of a pair of op-amps is shown in figure 24.9.
   Some of the resistor values are missing.
   a) State the type of amplifier that OA_1 and OA_2 have been wired up as.
   b) The inputs D, C, B and A can be at $+5$ V or 0 V. If V_{OUT} is to be $+0.8$ V when DCBA $= 0100$, work out the value of R_F.
   c) The value of V_{OUT} is supposed to be proportional to the value of the binary number DCBA. What are the values of R_D, R_B and R_A?
   d) Copy and complete the table below.

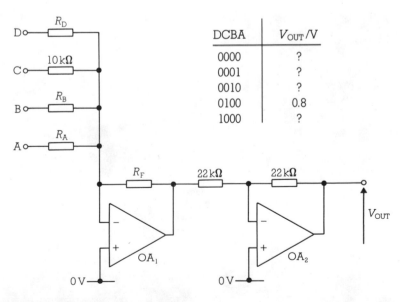

DCBA	V_{OUT}/V
0000	?
0001	?
0010	?
0100	0.8
1000	?

Figure 24.9 Question 3

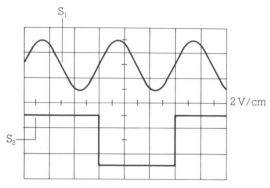

Figure 24.10 Question 4

4 Figure 24.10 shows two waveforms displayed on the screen of a double beam oscilloscope. Each signal has an amplitude of 2 V. They are fed into the audio mixer of figure 24.6. Draw the waveform which would appear on the screen of an oscilloscope connected to the output of the audio mixer.

5 A recording studio needs a four channel audio mixer, with a separate volume control for each channel. Draw a suitable circuit, with the component values clearly shown. Make sure that the op-amp does not saturate if all four signals to be mixed have amplitudes of less than 4 V.

6 DACs are used to allow microprocessor systems to generate complex sounds.
   a) With the aid of a block diagram, explain how the sounds are generated.
   b) Suggest, with reasons, a useful application for each of the following portable electronic systems.
      i) A talking thermometer. ii) A talking compass.
      iii) A talking calculator.

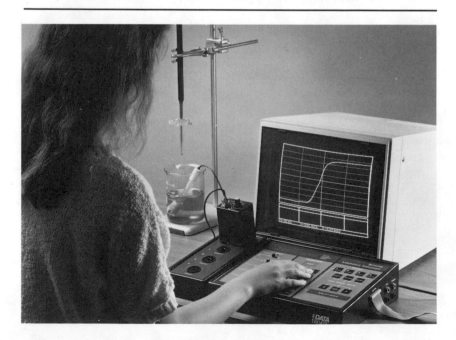

The pH probe in the beaker generates an analogue signal. That signal is converted to digital form by an ADC so that it can be processed by a microprocessor and stored in RAM. The TV screen shows how the signal from the probe has changed with time. *(Educational Electronics)*

25
Analogue-to-digital converters

This chapter is going to show you how analogue signals can be converted into digital form. That is, how to generate a binary word which represents a particular voltage. Devices which can perform this task are called **analogue-to-digital converters** or **ADCs**.

The need for ADCs

Electronics is about the gathering, processing, storing and transmission of information.

Information gathering tends to involve analogue electronics. For example, to gather information about temperature you would use the circuit of figure 25.1. Its output would be a voltage whose value would depend on the temperature of the thermistor.

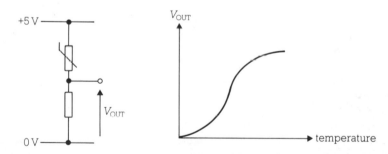

Figure 25.1 A thermistor generating a temperature dependent signal

The processing and storing of information tends to involve digital electronics. Logic systems and memories process and store information in the form of binary words. In particular, microprocessor systems can *only* handle information in digital form.

So there is a need for a device which can convert information in analogue form into digital form.

Using an ADC

Figure 25.2 shows an example of an ADC in action. The whole system records the temperature of the thermistor over a 24 hour period. The ADC produces eight bit words which represent the voltage fed out of the voltage divider. Those binary words are stored in the RAM.

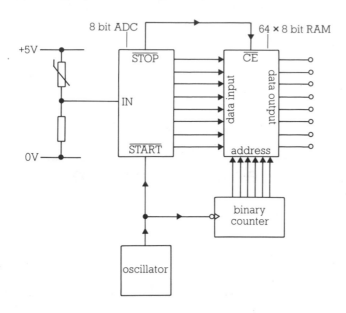

Figure 25.2 Using an ADC to store analogue signals in a RAM

The oscillator produces one pulse per half hour. That pulse enters a binary counter whose outputs provide the RAM address. So every half an hour the address goes up by one. At the same time, the pulse enters the $\overline{\text{START}}$ terminal of the ADC. This triggers the ADC. A short while later, $\overline{\text{STOP}}$ goes low and the ADC feeds the eight bit word into the RAM.

At the end of 24 hours the forty eight words stored in the RAM can be transferred into some other device, such as a microcomputer.

Inside an ADC

There are several ways in which analogue signals may be converted into digital form. The slowest, and least reliable, involves using a human being to look at a voltmeter, convert the reading into a binary number and type it into a keyboard ! The fastest, and most expensive uses long chains of op-amps (rather like the system of figure 14.10). The system which we are going to describe is relatively simple, very precise, cheap and moderately fast. It is called a **slope conversion ADC**.

A four bit ADC

The block diagram of a four bit ADC is shown in figure 25.3. The analogue signal to be converted is V_{IN}. When the system is triggered by pulling $\overline{\text{START}}$ low, there is a delay of a few milliseconds before the four bit word DCBA is ready at the digital outputs.

How it works

The timing diagram shows what happens when the system is triggered.

The ADC is triggered by a falling edge fed into the $\overline{\text{START}}$ terminal. $\overline{\text{START}}$ goes to the reset terminal of the binary counter, so when it goes low the counter starts to count the 1 kHz pulses from the oscillator. So DCBA goes up by one each millisecond.

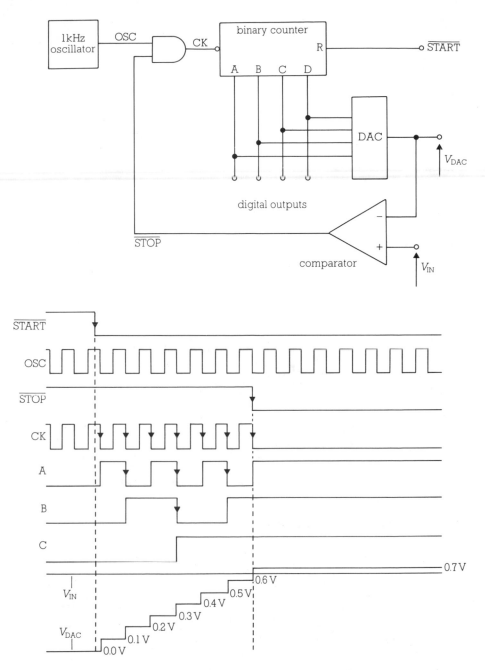

Figure 25.3 A four bit analogue-to-digital converter

The binary counter feeds a four bit word into a DAC. The DAC is of the type shown in figure 24.1, so its output (V_{DAC}) goes up by 0.1 V/ms. V_{DAC} **ramps** up.

V_{DAC} and V_{IN} are fed into a **comparator**. Provided that V_{IN} is higher than V_{DAC}, $\overline{STOP}$ will be high. As soon as V_{DAC} ramps above V_{IN} $\overline{STOP}$ will go low.

$\overline{STOP}$ controls the flow of pulses between the oscillator and the binary counter. As soon as $\overline{STOP}$ goes low, CK is forced to be 0 and the word DCBA is frozen.

The timing diagram has V_{IN} equal to $+0.65$ V. So when $\overline{STOP}$ goes low, DCBA is 0111. The binary output tells you, within 0.1 V, the value of the analogue input.

More bits, more speed

The **conversion time** of the ADC of figure 25.3 is between 1 and 15 ms. This could be speeded up by increasing the frequency of the oscillator. So a 100 kHz oscillator would give a conversion time of less than 0.15 ms. This is about as fast as CMOS logic gates can function reliably. If a different make of IC called TTL is used, conversion times of 1.5 μs can be attained.

The **precision** of the ADC is 0.1 V over a **range** of 1.5 V. By using an eight bit DAC instead of a four bit one, the precision can be reduced to 0.006 V for the same range. This will, of course, increase the conversion time to 255 ms. So the gain in precision is at the expense of a longer conversion time.

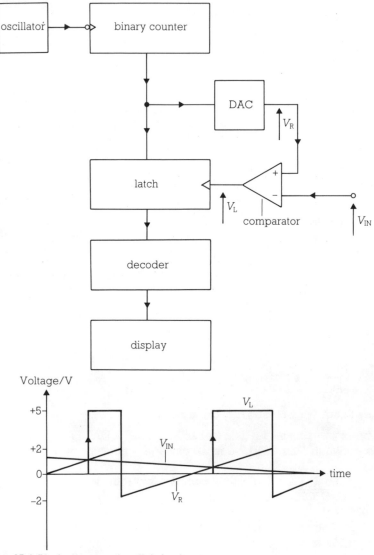

Figure 25.4 Block diagram of a digital voltmeter

Digital voltmeters

ADCs are widely used in measuring instruments which have digital displays. Multimeters, thermometers, weighing machines and micrometers are all examples of devices which benefit by having digital displays.

A digital display gives more precision, is quicker to read and is less likely to be misread. Furthermore, it tends to be more rugged than its analogue equivalent (a moving-coil meter). Finally, it can be cheaper and smaller than alternative displays. All of the devices mentioned above are basically **digital voltmeters** (or **DVM**s). They continuously sample an analogue signal, convert it into digital form and show its value via a seven segment display.

A DVM

The block diagram of the important parts of a DVM is shown in figure 25.4. It is basically an ADC which samples the analogue input signal V_{IN} at regular intervals and uses its digital output to run a seven segment display. The display typically has four digits, reading from $+2000$ to -2000. So the ADC requires a precision of 1 part in 4000. Since $2^{12} = 4096$, the ADC will have to feed out a twelve bit word.

How it works

The oscillator continuously feeds pulses into the twelve bit counter. The oscillator runs at a high frequency so that the counter counts from 0 to 4095 several times a second.

A twelve bit DAC uses the counter outputs to generate a ramp voltage V_R. V_R ramps up from -2.048 V to $+2.047$ V in steps of 0.001 V.

As soon as V_R has a higher value than V_{IN} the output of the comparator rises from 0 to 1. That rising edge triggers a twelve bit latch so that the word going into the DAC at that instant is stored.

The twelve bit word stored by the latch is passed onto a decoder and four digit display. The data shown on the display will be updated several times a second, each time that V_R rises above V_{IN}.

The behaviour of the whole system is summarised in the graphs of figure 25.4.

QUESTIONS

1 Explain what an ADC is supposed to do.

2 The three bit DAC of figure 25.5 has the characteristics shown in the margin.

Binary input CBA	Analogue output/mV
000	0
001	100
010	200
011	300
100	400
101	500
110	600
111	700

a) Plot a graph to show how the output of the DAC (V_{DAC}) changes with time.

b) What does the seven segment LED display as time goes on?

c) Copy the diagram of figure 25.5 inserting the following items to convert it into a simple digital voltmeter:
   comparator, three bit latch.

d) Explain how your digital voltmeter works.

e) What will the seven segment display show if V_{IN} is
   i) $+250$ mV,
   ii) $+520$ mV,
   iii) $+950$ mV?

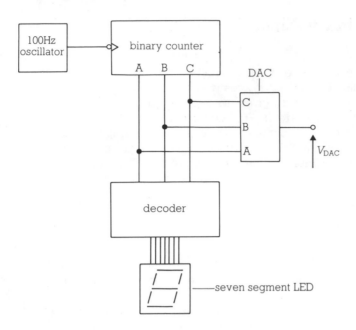

Figure 25.5 Question 2

3   A comparator behaves like an op-amp at its input, but like a logic gate at its output. The type of comparator used in figure 25.3 has the following characteristics.

Input signals	Output signal/V
$V_+ > V_-$	$+4.5$
$V_+ < V_-$	$+0.5$

Figure 25.6 shows a circuit with these characteristics. The op-amp is run off a $+5$ V and -5 V power supply.
a) When V_+ is higher than V_-, what is the voltage at
   i)   A,
   ii) OUT?
b) When V_+ is lower than V_-, what is the voltage at
   i)   A,
   ii) OUT?
Figure 25.7 shows another way in which a comparator can be constructed. The diode and 10 kΩ resistor make the output signal compatible with CMOS logic gates.
c) Describe the charateristics of this comparator with a table like the one above.

4   State, with reasons, in what ways digital voltmeters are better than moving-coil voltmeters.

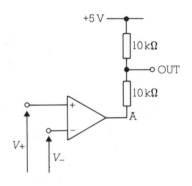

Figure 25.6 Question 3

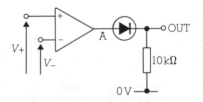

Figure 25.7 Question 3

Revision questions for Section E

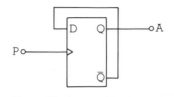

Figure E.1 Question 1

1 The D flip-flop of figure E.1 is triggered by rising edges.
  a) If A is 1, what does it become if P is
    i) lowered from 1 to 0,
    ii) raised from 0 to 1?
  b) What do you have to do to the input P to make the output A change state?
  c) Draw a circuit diagram to show how two D flip-flops can be made into a two bit binary counter. Label the outputs of the counter B and A.

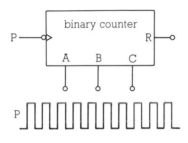

Figure E.2 Question 2

2 A three bit counter is shown in figure E.2. It is reset when R is held at 1.
  a) What is the value of the word CBA when R is 1?
  b) Three pulses are fed into P while R is held at 1. What happens to the word CBA?
  c) If CBA is initially 000 and R is held at 0, how many pulses have to be fed into P before CBA is 101?
  d) Copy and complete the table below. It shows what happens to CBA as pulses are fed in at P while R is 0.

Pulse number	C	B	A
0	1	0	0
1	?	?	?
2	?	?	?
3	?	?	?
4	?	?	?

  e) Figure E.2 shows the first line of a timing diagram. Copy it. Draw waveforms for A, B and C underneath it. Assume that CBA is initially 000 and that R is held at 0 throughout.

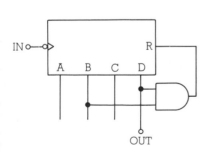

Figure E.3 Question 3

3 The circuit of E.3 feeds out a pulse at OUT for every ten pulses that are fed into IN.
  a) Initially DCBA is 0000. Draw up a table to show what happens to DCBA as ten pulses are fed into IN.
  b) Draw a timing diagram to show how OUT is related to IN. Assume that IN is a square wave; draw twelve cycles of its waveform.
  c) Show how you would connect two of the circuits of figure E.3 to make a system which would feed out one pulse for every hundred pulses fed in.
  d) Show how you would adapt the circuit of figure E.3 so that it would feed out one pulse for every six pulses fed into it.

4 The counter system shown in figure E.3 counts from 0 to 9. You have to design a system which will count the number of times that a switch is pressed and display that number on a seven segment LED. Draw a circuit diagram of the system. Explain what each part of the system does.

5 A sequence generator is shown in figure E.4. The oscillator feeds out one pulse every ten seconds.

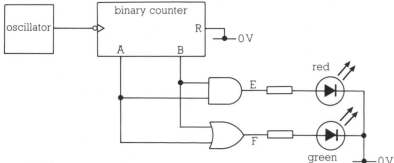

Figure E.4 Question 5

a) Draw up a truth table to show how E and F are related to B and A.
b) At a particular instant, both LEDs go off. Describe what each LED does during the next minute.

6 It is often claimed that electronically controlled machinery can produce articles better than human beings can. Describe the sort of tasks at which electronic systems are better than human ones. Explain why electronically controlled machinery may be more economical to run than a human labour force.

7 Figure E.5 is a block diagram of a simple microprocessor system.

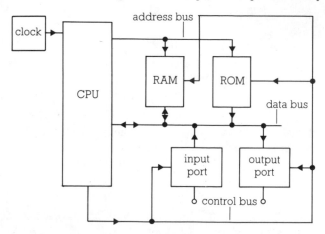

Figure E.5 Question 7

a) Describe the function of the ROM in the system.
b) What is the function of the CPU?
c) Which block is made up of tristates ? What is its function?
d) Which block is made up of D flip-flops? What is its function?
e) Explain what the data and address buses are and what they are used for?

8 Microprocessor systems are programmable. What does this mean?

9 With the help of a block diagram, explain how the connection of a digital-to-analogue converter allows a microprocessor system to produce sounds. Describe two useful applications of such a system.

10 Figure E.6 shows a three bit digital-to-analogue converter made out of two op-amps. The inputs (C, B and A) are either at +5 V or 0 V.

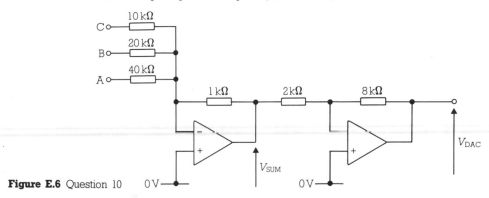

Figure E.6 Question 10   0 V

Calculate the value of V_{SUM} and V_{DAC} when CBA is
a) 000,   b) 001,   c) 010,   d) 100.

11 A three bit analogue-to-digital converter (the MWB0300) is shown in figure E.7. It has the characteristics shown in the table below, and is triggered by holding $\overline{START}$ low.

Input voltage/V	C	B	A
below 0.0	0	0	0
1.0 – 0.0	0	0	1
2.0 – 1.0	0	1	0
3.0 – 2.0	0	1	1
4.0 – 3.0	1	0	0
5.0 – 4.0	1	0	1
6.0 – 5.0	1	1	0
above 6.0	1	1	1

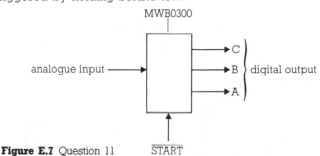

Figure E.7 Question 11 $\overline{START}$

a) Explain what an analogue-to-digital converter is supposed to do.
b) What does the word CBA become shortly after $\overline{START}$ is held low if V_{IN} is
   i)   +2.5 V,   ii)   +8.2 V,   iii)   +0.7 V?
c) The system of figure E.7 can be built around the system of figure E.6. Draw a block diagram to show how it can be done, using the following blocks; three bit DAC, binary counter, oscillator, comparator, AND gate.
d) Explain how the system you drew in c) works.

12 A microprocessor system with an analogue-to-digital converter connected to its input port can be used to recognise human speech.
a) Explain why speech recognition systems are expensive when they first appear, but become considerably cheaper after a while.
b) Describe, with reasons, a useful application of a speech recognition system which would take place while they were expensive.
c) Describe another useful application of speech recognition systems which would only happen when they were cheap.
d) In what ways could speech recognition systems help the disabled?

Wireless communications

A satellite communications receiver dish in West Africa. Good communications help developing countries. Some satellites can provide warnings of major changes in the weather; this allows planting or harvesting to be carried out at the best time. (Science Photo Library)

26
Radio

This chapter is about **wireless communication**. It will show you how electronics can be used to send information from one place to another with the help of **radio waves**.

Radio

A block diagram of a system which uses radio waves to transmit information is shown in figure 26.1. The **transmitter** is on the left and the **receiver** is on the right. The system can be used to send Morse code signals from one place to another.

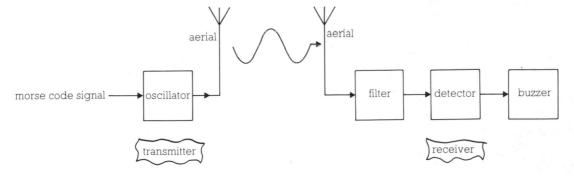

Figure 26.1 A Morse code wireless transmission system

The transmitter consists of an **oscillator** and an **aerial**. The oscillator makes alternating currents go up and down the aerial. This generates **radio waves** which travel away from the transmitter at the speed of light. The Morse code signal turns the oscillator on and off, so radio waves are emitted from the aerial in bursts.

If any of the radio waves hit the receiver aerial they generate alternating currents in it. The aerial signal passes through a **filter** before it is **detected**. The output of the detector goes high whenever a radio wave of the right frequency goes past the aerial. So as the bursts of radio waves pass the aerial, the buzzer goes on and off.

Radio waves

Radio waves are generated whenever an alternating current goes along a length of wire.

For example, if an alternating voltage of the correct frequency is fed into a **dipole aerial** (figure 26.2) radio waves with that frequency are produced. A dipole aerial emits radio waves efficiently if it is half a wavelength long. The wavelength of a radio wave is given by this formula;

$$\textbf{wavelength (m)} = \frac{\textbf{speed of light } (3 \times 10^8)}{\textbf{frequency (Hz)}}$$

The radio waves used to carry TV signals have a frequency of about 500 MHz (**1 MHz (megahertz) = 1000 kHz**). So a dipole used for TV transmissions needs to be about 30 cm long.

Radio waves have a **polarisation**. This is dictated by the orientation of the aerial. For example, TV transmitters often have dipoles which are horizontal, so that their radio waves are horizontally polarised.

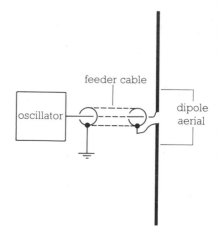

Figure 26.2 Generating radio waves with an oscillator and an aerial

Receiving aerials

A dipole aerial can also be used to receive radio waves. For efficient reception the dipole has to be half a wavelength long and correctly orientated to match the radio wave's polarisation.

Since the radio waves emitted by a transmitting aerial go off in all directions they get progressively weaker as they travel. Usually they are quite weak by the time they pass the receiver aerial, so the alternating currents induced in it will be small. Some form of **reflector** is often placed behind the aerial to increase the strength of the radio wave which hits it.

Frequency bands

Radio waves used for communications have frequencies between 30 kHz and 30 GHz. (**1 GHz (gigahertz) = 1000 MHz.**) They are divided into six frequency bands, shown in the table below. Each band is best suited to particular uses, some of which have been listed.

Frequency band	Uses	Wavelength
30 kHz		10 km
Long Wave	Long distance communication	
300 kHz		1 km
Medium wave	Local AM broadcasting	
3 MHz		100 m
Short wave	CB and amateur radio	
30 MHz		10 m
VHF	Local FM broadcasting and police	
300 MHz		1 m
UHF	TV broadcasting	
3 GHz		10 cm
Microwaves	Telephone links Satellite communication Radar	
300 GHz		1 cm

Range

The effective range of a radio signal depends on its power and its frequency. A high power transmitter will be detectable at greater distances than a low power one will.

Long wave signals have no difficulty in following the curvature of the earth's surface, so they can be used for long distance communication. On the other hand, the very short wavelength signals of the VHF (**v**ery **h**igh **f**requency) and UHF (**u**ltra **h**igh **f**requency) bands can only travel in straight lines. This means that they cannot be used for **over-the-horizon** communications. To maximise their range, UHF and VHF transmitters are located on high ground at the top of high masts.

Short wave radio signals cannot follow the curvature of the earth's surface very well. Yet they can be used for communicating with the other side of the earth! This is because they are reflected by the **Heaviside layer**, a band of ionised air in the upper atmosphere. Its height varies during the day and night as the atmosphere heats up and cools down, so the range of a short wave signal can depend on the time of day.

Satellites

Microwave signals have very short wavelengths. Their transmitting dipoles are less than 50 mm long, so it is possible to place reflectors behind them to beam the radio waves in one direction. If the wave is directed towards a **communications satellite** placed in orbit above the earth, the satellite can re-emit the wave so that it can be received over a large area of the earth's surface. Although the power of the satellite's transmission is inevitably low (it has to use solar cells to convert sunlight into electricity), large reflectors can be placed behind the receiving aerials on earth to compensate.

The arrangement of a satellite communication system is shown in figure 26.3. The satellite is usually placed in orbit so that it rotates at the same speed as the earth, staying above the same place all the time. (This is

A communications satellite. Note the reflecting dishes placed behind the receiving and transmitting dipoles. The large 'wings' hold the solar cells which provide the satellite with electrical energy.

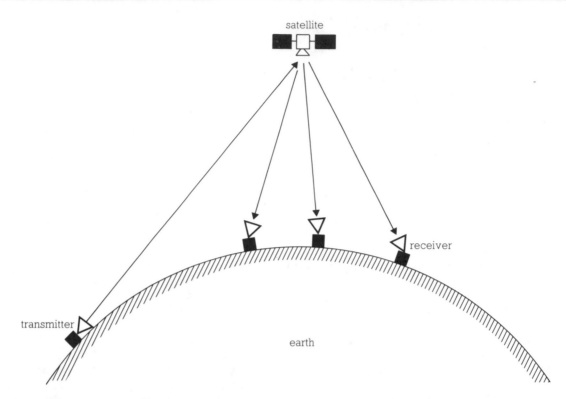

Figure 26.3 Using a communications satellite to send information around the curved surface of the Earth.

called a **geostationary orbit.**) A single satellite can, for example, broadcast TV signals to a whole continent. Despite the expense of building the satellite and placing it in orbit, it is usually a far cheaper alternative to a network of receivers and transmitters which can see each other over the curved surface of the earth.

Use of frequencies

The use of radio waves is very carefully controlled by international agreement. It is vital that two stations do not broadcast signals at the same frequency otherwise they could not be unscrambled by the receiver! So each frequency band is divided up into a number of **channels**, and each channel is allocated to a specific use. **It is illegal to transmit radio waves without proper authorisation, such as a licence.**

Broadcasting stations which are long distances apart can share channels as there is little chance that their signals will have a long enough range for them to interfere with each other. Thus the UHF band can have a number of different users allocated to a particular channel. Only when atmospheric conditions are particularly unfavourable will signals from an over-the-horizon transmitter be strong enough to interfere with signals from the local transmitter.

The demand for channels is high, so strict limits are placed on the power of transmitters. This deliberately limits the range of their signals, so that users who are not over the horizon can use the same channel. For example, the power of citizen band (CB) transmitters is strictly regulated so that they can only be used for local communications.

Modulation

Modulation is the name given to the process by which an oscillating waveform is made to carry some information. A waveform can be modulated in two different ways.

AM

The Morse code system of figure 26.1 uses **amplitude modulation**. The dots and dashes which make up the signal are used to switch the radio wave (called the **carrier**) on and off. This is shown in figure 26.4. The amplitude of the carrier (zero or non-zero) contains the information about the signal.

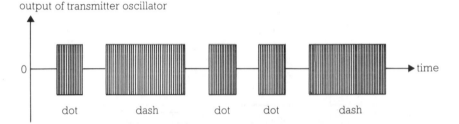

Figure 26.4 An amplitude modulated carrier

FM

Frequency modulation uses the signal to alter the frequency of the carrier. For example, frequency modulation is used to send digital signals down telephone wires. A frequency of 2.4 kHz represents a 1 and 1.2 kHz represents a 0. This is shown in figure 26.5. The frequency of the carrier contains the information about the signal.

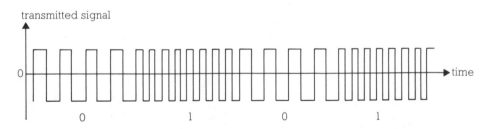

Figure 26.5 A frequency modulated carrier

Audio signals

Both amplitude modulation (**AM**) and frequency modulation (**FM**) can be used to modulate audio signals (such as speech or music) onto radio waves.

AM broadcasts are made in the long and medium wave bands, with FM broadcasts restricted to the VHF bands. This is because FM radio waves which carry audio signals have to be more than 100 kHz apart, whereas

AM radio waves need only be 10 kHz apart. Frequency modulation requires a much larger **bandwidth** (spread of frequencies) than amplitude modulation. On the other hand, FM signals are less susceptible to interference so they are used for high fidelity broadcasts.

AM transmitters

The block diagram of an AM transmitter is shown in figure 26.6. Its operation is quite easy to understand. The audio signal is used to wobble one of the radio frequency oscillator's supply rails up and down. The result is a high frequency signal whose amplitude goes up and down in time with the audio signal (figure 26.7).

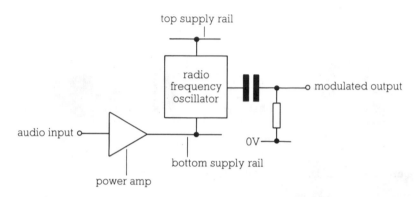

Figure 26.6 An amplitude modulator

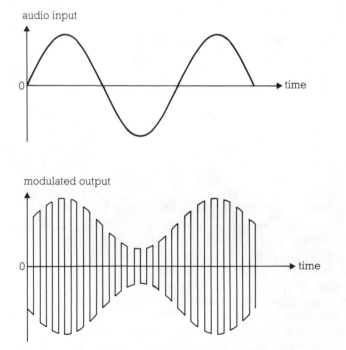

Figure 26.7 Input and output waveforms of an amplitude modulator

Radio receivers

The techniques which are used to receive radio waves and extract information from them depend on the frequency of the radio waves and how they have been modulated. The rest of this chapter will deal with systems which pick up signals in the long wave and medium wave bands where amplitude modulation is used.

At higher frequencies (short wave, VHF and UHF), different methods have to be used because of the difficulty of building high frequency amplifiers. Those methods (called **superheterodyning**) are beyond the scope of this book. High frequency circuits tend to use discrete components rather than integrated circuits and component layout becomes very important. Each part of the circuit needs to be carefully screened from the rest with aluminium or copper sheets. At very high frequencies (the microwave band) electronics looks like plumbing, with copper pipes to carry the signals between components!

AM receivers

The block diagram of a typical medium wave radio receiver is shown in figure 26.8.

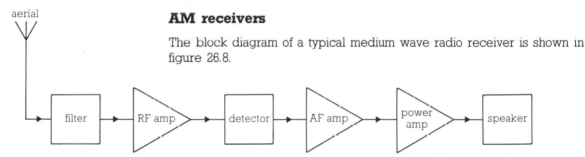

Figure 26.8 Block diagram of a MW radio receiver

The **aerial** picks up any radio waves which pass it, so a number of alternating currents at different frequencies are fed into the **filter**. The filter rejects all of those signals except one, so a single alternating voltage is fed into the radio frequency (**RF**) amplifier.

The amplified signal is then **detected** or **demodulated**. This extracts the audio frequency (**AF**) signal which was used to amplitude modulate the radio carrier.

Finally, the AF signal is further amplified and fed into a loudspeaker which generates sound.

Tuning

The circuit shown in figure 26.9 is the usual method of selecting one out of the many radio signals picked up by an aerial.

The aerial feeds alternating currents into an **inductor** connected in parallel with a **variable capacitor**. This combination makes a **tuned circuit**, a filter which lets all alternating currents through to ground except one. Alternating currents with a particular frequency (the **resonant frequency** of the circuit) generate an alternating voltage at the output of the circuit.

The resonant frequency of the tuned circuit depends on the values of the inductor and the capacitor. By adjusting the variable capacitor, the resonant frequency can be made the same as that of the radio wave you want to receive. As that wave makes alternating currents go up and down the aerial a small alternating waveform will appear at the output.

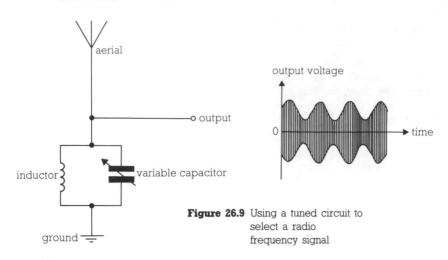

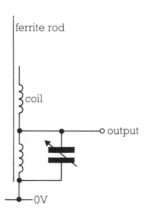

Figure 26.9 Using a tuned circuit to select a radio frequency signal

Aerials

The circuit of figure 26.9 will only work well if the aerial is connected to ground via the tuned circuit. Dipole aerials are not feasible for long wave and medium wave receivers (they would have to be at least 50 m high!) so a vertical length of wire has to do instead. Portable receivers often use a **ferrite rod** aerial. This is a bar of ceramic which contains iron. It usually has the inductor (which is a coil of wire) wound around it. The ferrite rod uses the alternating magnetic field of the radio wave to induce an alternating voltage across the ends of the inductor. A typical arrangement is shown in figure 26.10.

Detection

The circuit shown in figure 26.11 can be used to extract the audio frequency signal which was used to modulate the amplitude of the radio frequency signal at the transmitter.

Figure 26.10 Using a ferrite rod instead of an aerial

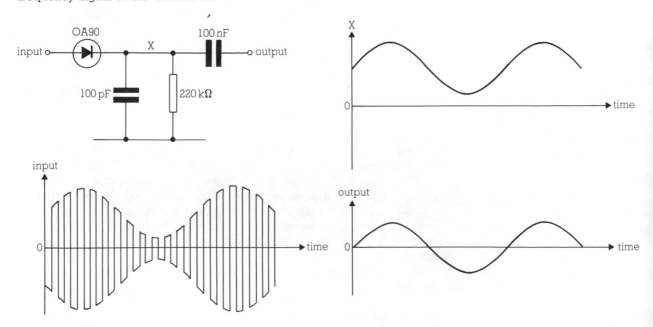

Figure 26.11 A detector for AM signals

A diode rectifies the RF signal. Then a 100 pF capacitor smoothes it. **(1000 pF (picofarad) = 1 nF.)** The 220 kΩ resistor allows the capacitor to discharge so that the voltage at X can follow the peak voltage of the input signal. The 100 nF capacitor provides AC coupling so that the output alternates with an average value of 0 V.

A special diode, known as a **point-contact germanium diode**, has to be used for the rectifying. This is because the high frequency of the RF signal means that the capacitance of the diode must be kept as low as possible. Furthermore, the amplitude of the RF signal is usually very small (less than 100 mV), so you have to use a diode with a low voltage drop between anode and cathode when it is forward biased.

Receiver circuit

The circuit diagram of a simple, but effective AM radio receiver is shown in figure 26.12.

The aerial is a vertical length of wire about two metres long. The values of the inductor and variable capacitor mean that the system can pick up radio waves whose frequency is roughly 1 MHz. It can, for example, receive Radio 1, Radio 2 and Radio 3.

The AF signal which comes out of the detector is amplified by the audio amplifier. The 10 μF capacitor AC couples that signal into a potentiometer which acts as a volume control.

Finally, a pair of transistors boost the power of the AF signal so that it can drive the speaker and generate a reasonable amount of sound.

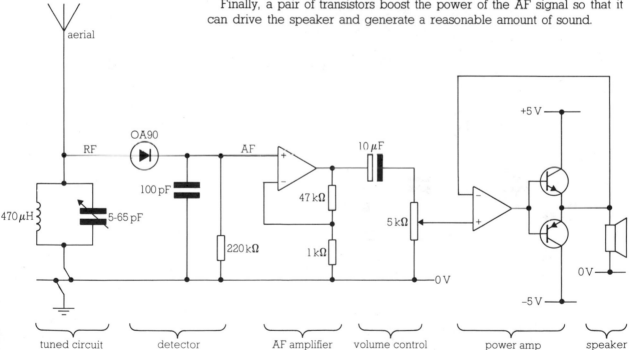

Figure 26.12 A MW radio receiver

QUESTIONS

1 State the range of radio frequencies which
   a) are used for broadcasting TV,
   b) are reflected by the Heaviside layer,
   c) provide local AM broadcasting,
   d) transmit FM radio.

2 Draw a block diagram of a radio wave transmission system which broadcasts speech using amplitude modulation. Use the following blocks; microphone, aerial, RF oscillator, AF amplifier, power amplifier. Explain the function of each block.

3 Draw a block diagram of a radio wave receiver system which plays back music which has been amplitude modulated onto a radio wave. Use the following blocks; aerial, speaker, power amplifier, RF amplifier, detector, AF amplifier, filter. Explain the function of each block.

4 What is the difference between amplitude modulation and frequency modulation ?

5 Figure 26.13 shows the circuit diagram of a very simple radio receiver. The headphones have a very high resistance so they do not require much current. (They don't generate much sound either!)

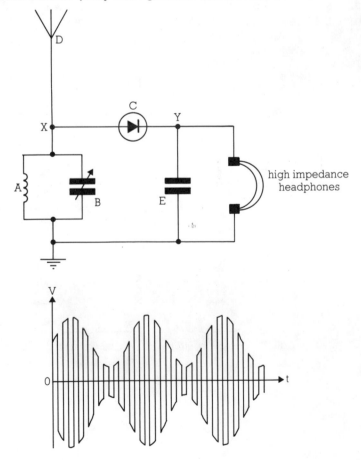

Figure 26.13 Question 5

a) Name the components A, B, C, D and E.
b) Which component is adjusted to tune into a radio station?
c) Copy and complete the following statements.
   When radio waves pass the .... they make .... currents flow in it. If the frequency of the current is the same as the .... of the .... circuit, an .... voltage is produced at X. This is .... by the diode and .... by the capacitor.

The graph shows how the voltage at X varies with time at one particular instant.

d) Draw a graph to show the voltage at Y during that instant.
e) Draw a graph to show what the voltage at Y would be if E was removed from the circuit.
f) Explain how you would modify the circuit of figure 26.13 so that it had a louder sound output.

6 Radio communications are extremely important for the police, ambulance and fire services. With the help of some examples, explain why.

7 Explain why a modern airline service could not operate without the use of radio communications.

27

Transistor amplifiers

Unless it is very near to the transmitter, the signal picked up by a radio aerial is very small. So that signal has to be amplified before you can detect it. But op-amps don't work at radio frequencies, so you have to use another sort of amplifier. This chapter is going to show you how transistors can be used to amplify high frequency signals. Transistor amplifiers can be very complicated, particularly when they are designed for very high frequencies. We shall stick to simple, but effective designs which use a single transistor.

The common-emitter amplifier

The transistor connection at the heart of a transistor amplifier is shown in figure 27.1 (a complete amplifier is shown in figure 27.4). It is known as the **common-emitter** connection because the emitter is held at 0 V.

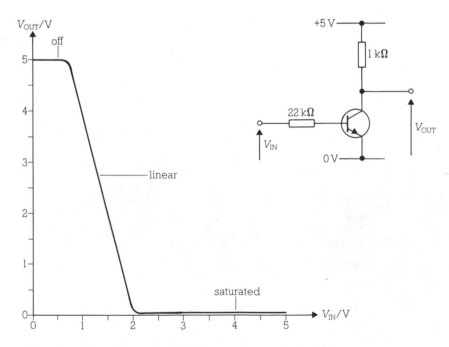

Figure 27.1 An npn transistor in the common-emitter connection

The transfer characteristic of the system is shown in figure 27.1. It has three regions.

1) When V_{IN} is below 0.7 V the transistor is **off**. No current goes into its collector so the 1 kΩ resistor pulls V_{OUT} up to 5 V.

2) When the transistor is **on** current will go into its collector. As that current goes through the 1 kΩ resistor it will bring V_{OUT} below 5 V. This is the **linear** region.

3) When V_{OUT} gets near to 0 V the transistor is **saturated**. The collector current is large, so V_{IN} must be fairly high to provide enough base current.

Gain

The graph of figure 27.1 shows that a small change of V_{IN} can cause a much larger change of V_{OUT}. As V_{IN} goes from about 1 V to 2 V, V_{OUT} drops from 5 V to 0 V.

This is just what we need for an amplifier. The system has a gain of five, because the change of V_{OUT} is five times larger than any change of V_{IN}.

Saturation

We can calculate exactly how big the gain is by finding out the value of V_{IN} which will just saturate the transistor. The symbols which will be used are shown in figure 27.2.

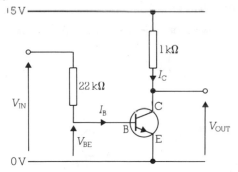

Figure 27.2 Current flow in the transistor

A transistor is saturated when its collector has the same voltage as its emitter. So V_{OUT} is 0 V. We can use Ohm's law to calculate the current going through the 1 kΩ resistor.

$$R = \frac{V}{I} \qquad \begin{array}{l} R = 1\text{ k}\Omega \\ V = 5\text{ V} \\ I = I_C \end{array} \qquad \text{therefore } 1 = \frac{5}{I_C}$$

$$\text{therefore } I_C = \frac{5}{1} = 5\text{ mA}$$

We are going to assume that the transistor has a current gain of 100. So we can use the value of the collector current to work out the value of the base current I_B.

$$I_C = h_{FE}\,I_B \qquad \begin{array}{l} I_C = 5\text{ mA} \\ h_{FE} = 100 \\ I_B = ? \end{array} \qquad \text{therefore } 5 = 100 \times I_B$$

$$\text{therefore } I_B = \frac{5}{100} = 0.05\text{ mA}$$

The base current has to go through the 22 kΩ resistor. So we can use Ohm's law to work out the voltage drop across it.

$$R = \frac{V}{I}$$
$$R = 22 \text{ k}\Omega$$
$$V = ?$$
$$I = 0.05 \text{ mA}$$

therefore $22 = \dfrac{V}{0.05}$

therefore $V = 22 \times 0.05 = 1.1 \text{ V}$

A transistor which is switched on has a voltage drop of 0.7 V between its base and its emitter. So V_{IN} is equal to the voltage drop across the 22 kΩ resistor plus 0.7 V. **V_{IN} must therefore be 1.8 V.**

We now know that if V_{IN} is 1.8 V then V_{OUT} will be 0 V. If V_{IN} is raised above 1.8 V the base current will rise, but the collector current will stay at 5 mA. Furthermore, if V_{IN} is below 0.7 V the transistor is off and V_{OUT} will be 5 V.

So when V_{IN} rises from 0.7 V to 1.8 V, V_{OUT} falls from 5 V to 0 V.

$$\text{Gain} = \frac{\text{output change}}{\text{input change}} = \frac{0 - 5}{1.8 - 0.7} = -4.5$$

Losing DC

The circuit of figure 27.2 cannot be used as an amplifier without adding more components. Although V_{OUT} will give out a big wiggle when a small wiggle is fed into V_{IN} that wiggle will not be centred on 0 V. So a **coupling capacitor** is needed to extract the amplified signal from the collector. This is shown in figure 27.3. Notice how any change in V_C causes an identical change in V_{OUT}, but that V_{OUT} is centred on 0 V.

Capacitors block DC signals but let AC signals through.

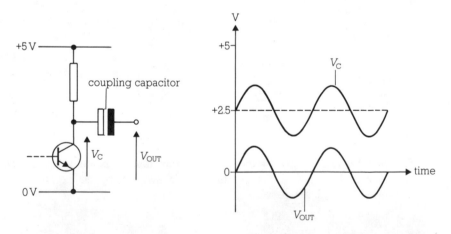

Figure 27.3 Using a coupling capacitor to extract an amplified signal from the collector

Bias

The full circuit of a common-emitter amplifier is shown in figure 27.4. The **bias network** on the left is a voltage divider which sets the value of V_{IN}. Adjustment of the potentiometer varies the amount of current fed into the base of the transistor.

**Correct setting of the bias network is crucial
if the amplifier is to work properly.**

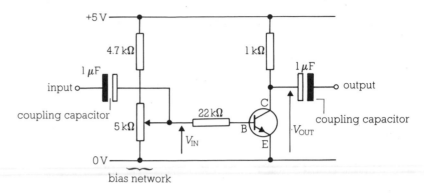

Figure 27.4 A common-emitter amplifier

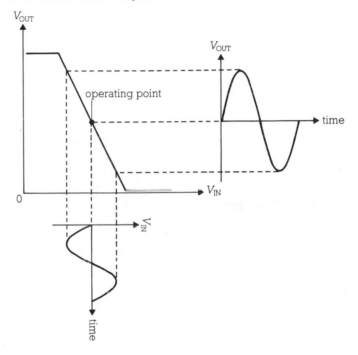

Figure 27.5 Input and output waveforms for correct bias

Figure 27.5 will help you to see why. It shows what happens when a small AC signal is fed into V_{IN} via a coupling capacitor. Provided that the **operating point** is halfway down the linear region of the characteristic, a larger but similar AC signal appears on V_{OUT}. Correct bias, with the average value of V_{OUT} lying halfway between the two supply rails, ensures that both the positive and negative parts of the AC signal are amplified without distortion.

Incorrect bias

Figure 27.6 shows what can happen if the bias network is **not** correctly adjusted. The operating point has been set too near the top of the linear region of the characteristic. So V_{OUT} cannot follow the changes in V_{IN} faithfully. The positive portions of the AC signal are amplified correctly but the negative portions are not. The output signal is distorted.

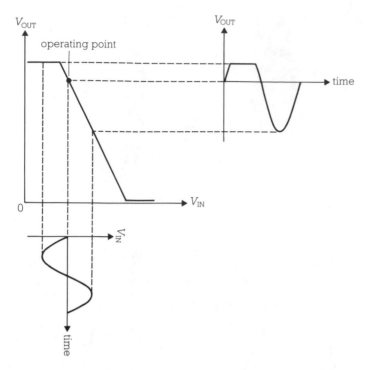

Figure 27.6 The consequences of incorrect bias

A better amplifier

Although the amplifier of figure 27.4 works, it is far from ideal. This is because both its bias adjustment and its gain depend on the transistor's current gain. Unfortunately, the current gain of a transistor is only approximately specified when you purchase it. Typically, it could be anywhere between 50 and 200. So every time that you assemble an amplifier you have to adjust its bias network and you only have a rough idea of what its gain is going to be.

This sort of circuit is bad news for the electronics industry. If you are building amplifiers you need to be able to specify exactly what their gain is going to be so that they all behave the same way. Furthermore, you

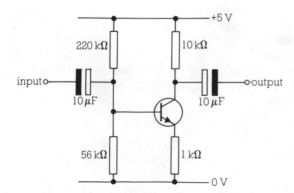

Figure 27.7 An amplifier which has a gain of -10 and needs no bias adjustment

don't want to have spend valuable time adjusting the bias network of each one. And the customer does not want to have to pay someone to re-adjust his amplifiers when their operating points drift away from the ideal setting. (The current gain of a transistor changes as it gets older.)

What industry needs is a design which needs no bias adjustment and whose gain is independent of the particular transistor used. The circuit of figure 27.7 goes some way towards providing that ideal. The collector will sit at 3 V, allowing V_{OUT} to have amplitudes of up to 2 V without distortion. The gain will be -10 provided that the transistor has a current gain of more than 100.

FET amplifiers

Transistor amplifiers can provide a reasonably big gain (up to 100) over a wide range of frequencies, but they suffer from one defect. Their input impedance is too small. This means that they can draw too much current from a signal source. In particular, they draw too much current from a tuned circuit which is feeding out RF signals. So another sort of RF amplifier is needed at the front end of radio recievers. Amplifiers based on **field effect transistors** (**FETs**) are suitable.

FETs

There are many types of transistor. The sort which you have been using in previous chapters is known as a **bipolar junction transistor**. The amplifier of figure 27.8 uses an **n-channel junction gate field effect transistor**.

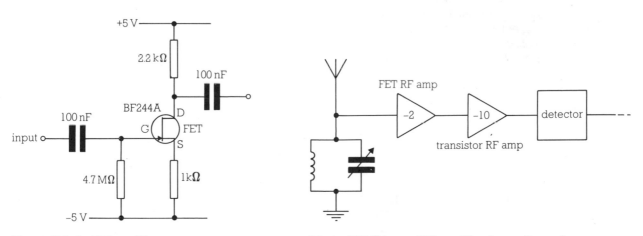

Figure 27.8 An FET amplifier

Figure 27.9 Using an FET amplifier in a radio receiver

Both types have three terminals, but that is just about all they have in common. They operate in completely different ways and have to be used in different arrangements to achieve the same function. So although the circuits of figures 27.7 and 27.8 are both amplifiers they have completely different bias arrangements.

**FETs make amplifiers with small gains
but very large input impedances.**

RF amplifiers

The amplifier shown in figure 27.8 has a gain of only two but has a very large input impedance of 4.7 MΩ. Figure 27.9 shows how such an amplifier might be used in a radio receiver. The FET amplifier takes the signal from the aerial and tuned circuit and amplifies it by -2. That can then be fed into a bipolar transistor amplifier with a larger gain of -10, giving a total gain of 20. Because the current flowing in the tuned circuit and aerial is very small it is imperative that the first amplifier takes virtually no current from it. FET amplifiers are often used to amplify signals from sources, such as human beings and chemical solutions, which cannot provide much current.

QUESTIONS

1　This question is about the circuit shown in figure 27.10. The transistor is saturated.
  a)　What is the value of V_{OUT}?
  b)　Calculate how much current goes through the 2.5 kΩ resistor.
  c)　The current gain of the transistor is 100. Calculate how large the base current must be.
  d)　Calculate the voltage drop across the 100 kΩ resistor.
  e)　How big must V_{IN} be?

2　Figure 27.11 shows how a transistor can be used to interface a logic gate (run off +5 V and 0 V) to a 12 V, 50 mA bulb. When Q is 1 the transistor must saturate so that the bulb comes full on. The current gain of the transistor is 100.
  a)　What will the bulb do when Q is 0?
  b)　When Q is 1, how much current must flow into
    i)　the collector,
    ii)　the base?
  c)　Should R be 10 kΩ or 4.7 kΩ? Explain your answer.

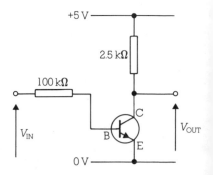

Figure 27.10 Question 1

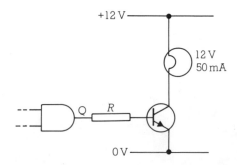

Figure 27.11 Question 2

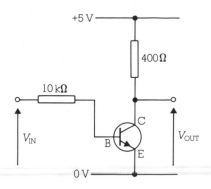

Figure 27.12 Question 3

3 The transistor shown in figure 27.12 has a current gain of 50.
   a) For each of the values of V_{IN} shown in the table below, calculate the following.
      i)   The base current.
      ii) The collector current.
      iii) The value of V_{OUT}.

V_{IN}/V	0.0	1.7	2.7	3.7	4.7
V_{OUT}/V	?	?	?	?	?

   b) Use the completed table to draw a transfer characteristic for the circuit.

4 Draw a circuit diagram to show how a transistor can be used to make an AC amplifier. Don't bother about component values. Indicate where signals are fed into the amplifier and where they come out of it.

5 Figure 27.13 shows voltage-time graphs of signals at four places in a transistor amplifier. All four refer to the same instant of time. State which of the waveforms [i), ii), iii) or iv)] belongs to A, B, C or D.

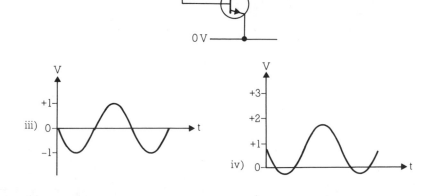

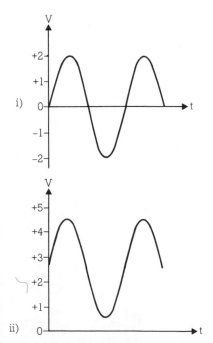

Figure 27.13 Question 6

6 Copy and complete the following statements. They apply to the circuit of figure 27.12.

When V_{IN} is 0 V the transistor is ..... Its base current will be ...., its collector current will be ..... and V_{OUT} will be ..... If V_{IN} goes above 0.7 V the transistor is ..... As V_{IN} is raised further V_{OUT} will ..... When V_{OUT} reaches 0 V the transistor will be ..... Any further increase of V_{IN} will increase the .... current but not change the .... current.

7 Draw the circuit diagram of an AC amplifier which uses an FET. Don't bother about component values. In what ways is its performance superior and inferior to a similar bipolar transistor amplifier? Suggest, with reasons, a good use for an FET amplifier.

28
Television

Chapter 26 showed you how radio waves are used to transmit sound from one place to another. This chapter will explain how black-and-white pictures are transmitted by radio waves.

Coding information

Radio waves can be used to transmit **any** information from one place to another. In the early days of wireless communication, radio waves carried information in the form of Morse code.

The message to be transmitted was first broken up into letters of the alphabet. Each letter was then given a sequence of dots and dashes according to the Morse code. The whole message therefore became a long string of dots, dashes and spaces which was used to turn the radio wave on and off.

The radio receiver would convert the bursts of radio waves which it recieved back into dots, dashes and spaces. Finally, the code would be used to convert that sequence into a string of letters of the alphabet, allowing the message to be reassembled.

All information is still transmitted this way.

Of course, the information needs to be suitably coded before it can modulate the radio wave. Furthermore, the receiver has to use the **same** coding for it to extract the information from the radio wave.

Morse code uses a series of dots and dashes. Speech and sound are coded as alternating voltages when they are transmitted on medium wave, long wave and FM transmissions. Computer data is coded as a string of 1's and 0's for transmission down telephone lines.

TV pictures are coded as AC signals which are amplitude modulated onto UHF radio waves.

Making pictures

Television pictures are made by varying the brightness of a large number of dots arranged in a rectangle (the screen). Each dot is called a **pixel**. They are arranged in 625 rows with 833 pixels in each row. So the whole picture is built up out of $625 \times 833 = 520\ 625$ dots whose brightness can

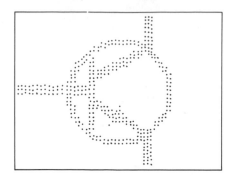

Figure 28.1 Making a picture out of
dots

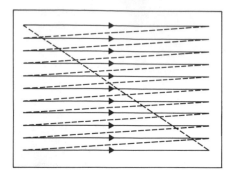

Figure 28.2 How line scans make
up a field scan

vary from black to white. This set of rows, or lines, which are used to make up the picture is called a **raster**.

Figure 28.1 shows how a picture can be made up of dots in this way.

Scanning

If you wanted to simultaneously control the brightness of each pixel in the raster you would need 520 625 different radio waves! You can, however, get away with using a single carrier wave if you **scan** the raster.

Rather than having all of the pixels lit all of the time, scanning involves making them glow one after the other in a sequence. That sequence is shown in figure 28.2.

Each row, or line, is scanned from left to right. The first row to be scanned is at the top of the picture and the last one is at the bottom. In this way a number of **line scans** are used to make a **field scan**.

Two field scans are needed to make up the whole raster. The first one lights up the pixels in the even numbered rows and the second one lights up the pixels in the odd numbered rows. This is called **interlacing**.

Scan rate

TV broadcasts in this country use a **field scan rate** of **50 Hz.** Since two field scans are needed to make all of the pixels in the raster glow, the whole picture is put on the screen in 1/25 s. Although each pixel glows for only about 0.1 μs, our eyes see the sequence of lit dots as a complete picture. The picture on the screen is renewed 25 times a second, but our eyes see it as being there all of the time.

Of course, if each field scan is slightly different from the previous one, the picture seen on the screen will appear to move.

Coded signals

Four separate pieces of information need to be coded for black-and-white TV transmission.

The most important one is the **video signal**. This tells each pixel how bright it is supposed to be. Since the pixels are lit one after the other the video signal is going to be an alternating voltage similar to that used for sound transmission. Its frequency range will be about 6 MHz, much larger than the 5 kHz used for sound transmissions.

The video signal is amplitude modulated onto the UHF radio wave. This

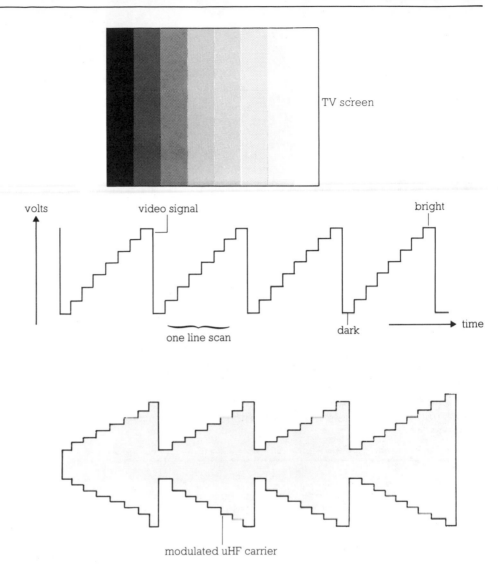

Figure 28.3 The video signal modulates the carrier and controls the brightness of pixels during a line scan

is illustrated in figure 28.3. Notice how the changes of amplitude of the modulated wave give rise to the pattern on the TV screen. A high amplitude makes the pixels glow brighter than a low amplitude does.

Synchronisation

Two **synchronising pulses** also have to be modulated onto the UHF carrier. These tell the receiver when to start a line scan and a field scan.

The **line sync pulses** have a frequency of 15 625 Hz. Each time one is detected by the receiver it starts a new line scan.

Field sync pulses have a frequency of 50 Hz and are much longer than the line sync ones. Each time that one of them is detected by the receiver a new field scan is started.

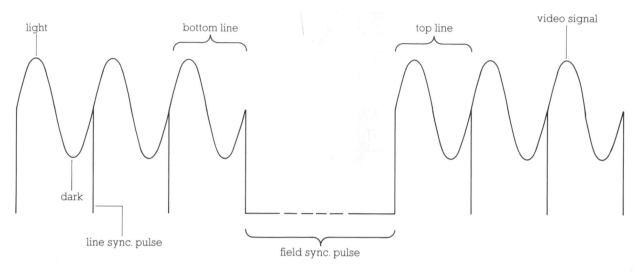

Figure 28.4 The line sync and field sync pulses added to the video signal

Figure 28.4 illustrates the final signal which modulates the UHF carrier. Notice how the line sync and field sync pulses **blank out** the video signal, with the field sync pulses lasting much longer than the line sync ones.

Sound

The **sound signal** which accompanies the video signal is frequency modulated onto a separate carrier which is 6 MHz higher than the UHF carrier.

QUESTIONS

1 Copy and complete the following statements.
   a) The picture on a TV screen is made up of . . . . . arranged in rows. The brightness of each pixel is controlled by the . . . . signal which is . . . . modulated onto the UHF carrier.
   b) Each time that a . . . . pulse is detected, the receiver starts a new line . . . . . A field sync pulse makes the receiver
       . . . .
   c) The field scan rate is . . . . Hz. Two field scans are interlaced to make a . . . . So the picture on the screen is renewed every . . . . s.

2 TV broadcasts are used for many things. Describe as many of them as you can.

3 TV cameras are widely used in the security business to keep an eye on what people are doing. Explain why this allows a few security guards to look after a large wharehouse.

4 TV cameras can act as eyes in places where it would be either too dangerous or inconvenient to send a human being. Describe an example of such a use in each of the following.
   a) Bomb disposal.
   b) Sewer inspection.
   c) Interplanetary exploration.
   d) Underwater mining.

29
Infra-red

Infra-red waves are radio waves which have a frequency of about 300 000 GHz. They behave like light waves (which have a frequency of about 500 000 GHz) but our eyes cannot see them. Their high frequency means that they can transfer information at a much higher rate than radio waves.

IR communications

The use of **optical fibres** to carry infra-red waves from one place to another allows enormous amounts of data to be transferred in a short time. Communication systems which use infra red and optical fibres are eventually going to replace many systems which use wires at the moment.

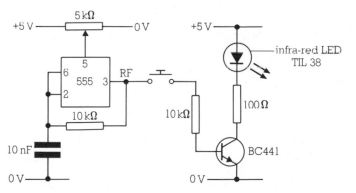

Figure 29.1 An IR transmitter

IR transmitters

A simple infra-red (**IR**) transmission system consists of an oscillator (the 555 IC), a transistor switch and an **infra-red LED** (see figure 29.1).

Each time that the switch is closed the transistor switches the LED on and off at 5 kHz. The whole system is effectively a Morse code transmitter. It emits short 5 kHz bursts of IR whenever the switch is closed.

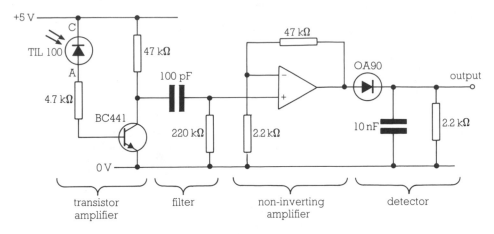

Figure 29.2 An IR receiver

IR receivers

A receiver which is compatible with the transmitter of figure 29.1 is shown in figure 29.2.

The pulses of IR which hit the **photodiode** allow small bursts of electric current to go through it. The photodiode is reverse biased so no current can normally go through it. Every time that the photodiode absorbs IR a small current can flow from cathode to anode. Since photodiodes can also respond to visible light they are often packaged in black plastic which is transparent to IR.

The bursts of current which flow through the photodiode are very small. So they are amplified by the transistor before being fed into a **filter**. This cuts out the 100 Hz IR signal which comes from the room lights, letting the 5 kHz photodiode signal through to the amplifier.

Finally, a diode **detector** rectifies the signal to generate the output signal. This will go high every time that a 5 kHz burst of IR is picked up by the photodiode. The output signal could, of course, drive a buzzer or a relay with the help of a suitable buffer.

Range

IR transmission systems are commonly used for the remote control of TV sets and music centres. A hand-held transmitter sends out IR signals which are picked up by the device being controlled.

They are ideally suited to a domestic environment because of their limited range. As the IR pulse moves away from the LED it becomes weaker as it spreads out. So the range of an IR signal is measured in metres rather than kilometres. The output power is much lower than that of the radio waves used for short distance communication. IR waves are strongly absorbed by walls, so they will not stray out of one room to interfere with similar systems in another room.

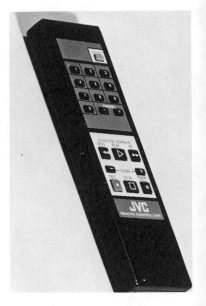

A portable IR transmitter. This one is used to control a video tape recorder.

Optical fibres

The range of an IR transmission system can be increased enormously if the pulses are fed down an **optical fibre**. This is a long thread of glass less than 1 mm across which can transmit IR pulses over long distances with very little loss. For example, a kilometre length of glass fibre may transmit 50% of the energy of an IR pulse. (A similar length of coaxial cable fed with 1 MHz pulses might only let through 0.1% of the signal!)

Talking down fibres

A block diagram of a two-way transmission system which uses optical fibres is shown in figure 29.3. It could be used to handle a two-way telephone conversation between two people.

Each of the conversations is converted to an alternating voltage by the microphones. These have to be coded into pulses before being transmitted down the optical fibre as bursts of IR. At the receiver the train of pulses is converted back into an alternating voltage which can generate sound waves via a speaker.

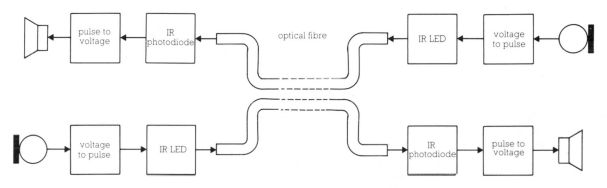

Figure 29.3 Using optical fibre to make a two-way telephone link

Interference

Although the two optical fibres will travel between the two locations along the same path, there is no way in which signals from one fibre can stray into another. So the system will have no **crosstalk** between the sending and receiving channels.

Furthermore, there is no way in which the pulses travelling down the fibres can suffer **interference** from objects outside. Unlike wire cables, optical fibres do not pick up mains hum or other signals from heavy electrical machinery. So optical fibres allow a low noise communication system to be built in environments which are electrically very noisy.

Security

For the same reason, optical fibres make a very secure communications system. People can listen in to radio broadcasts, they can tap telephone cables, but they cannot extract pulses from an optical fibre without cutting it in half. So it is virtually impossible for someone to listen in undetected to a conversation transmitted by an optical fibre!

Fibres for telephones

Optical fibres can transmit information at a prodigious rate. They can transmit up to 100 million pulses per second over short distances (1 km). A telephone conversation might need only 70 000 pulses per second for transmission by an IR system. Clearly, a single optical fibre should be capable of simultaneously transmitting several hundred separate telephone conversations.

In other words, a single optical fibre can replace a hundred separate wire telephone cables!

Multiplexing

The block diagram of figure 29.4 shows how several separate signals can be **multiplexed** onto a single optical fibre.

A number of telephone signals are fed in at the left. Each will be an alternating voltage with a frequency between 300 Hz and 3.4 kHz. They are each sampled 6800 times a second by an analogue-to-digital converter. This feeds out an eight bit word which represents the instantaneous voltage of the telephone signal.

The outputs of the ADCs are fed, in turn, by a switching network down the optical fibre as a series of pulses.

At the receiving end the pulses are used to recreate the eight bit binary words. These are fed into ADCs to generate the alternating voltages of the original telephone signals.

So the voltage of each telephone line at the right hand end of the system is updated 6800 times a second. Since each binary word needs about 10 pulses to be transmitted, each separate conversation involves about 70 000 pulses per second down the optical fibre. So a hundred separate simultaneous telephone conversations would only need about 10 million pulses per second, including synchronising pulses.

Of course, this system requires a large amount of complex circuitry at the transmitting and receiving ends of the fibre. However, for long distance communications, the electronics is comparatively cheap compared with the cost of the link itself. A single link is far cheaper than a hundred links!

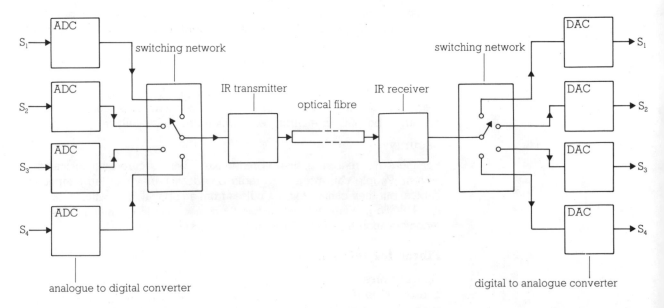

Figure 29.4 Multiplexing analogue signals onto a single optical fibre

Fibre versus cable

Coaxial cable can also be used for multiplexing signals in the same way. Indeed, the majority of such signals are currently routed via coaxial cable. But optical fibres are soon going to replace coaxial cable in many applications, even though they cannot transmit more pulses per second.

Optical fibres are far smaller and lighter than coaxial cable. So a single cable can be replaced by a number of fibres without the need to enlarge the ducts that they sit in.

Optical fibres have less loss. A coaxial cable handling 10 million pulses per second might need an amplifier every 0.5 km to boost the signal. These amplifier are called **repeaters** and they are a nuisance. They need to be supplied with power and are often stuck in inaccessible places (like under the sea!). An optical fibre will only need a repeater every 15 km, giving it a significant advantage.

Optical fibres are less susceptible to interference. So they permit higher quality transmissions than coaxial cables do.

QUESTIONS

1   Figure 29.5 shows a simple alarm system. The circuit on the left is portable. Each time that its switch is pressed the buzzer is supposed to make a noise.
   a)   Name component X. What does it do when the switch is pressed?
   b)   Name component Y. What happens to the current going through it when the IR hitting it increases?
   c)   For Q to go high, C must go below 1.3 V. Assuming that the current gain of the transistor is 200, calculate the following for C at 1.0 V.
       i)   the voltage drop across the 220 kΩ resistor,
       ii)   the collector current,
       iii) the base current.

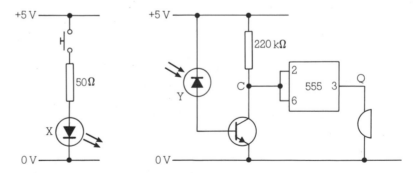

Figure 29.5 Question 1

2   Devise a burglar alarm system which uses an IR detection system as follows. An LED transmits IR across the room to a photodiode. Provided that IR is being absorbed by the photodiode the alarm stays silent. If a burglar blocks off the IR for an instant the alarm comes on and stays on.
   a)   Draw a block diagram of the system. Use it to explain how the system works.

b) Draw a circuit diagram, without bothering about component values.

3   IR transmission systems are widely used for the remote control of devices. Describe how such systems could be useful for disabled people who are not able to move easily around their homes.

4   State the ways in which optical fibres are superior to coaxial cables for transmission systems.

5   Draw the block diagram of a system which would open a door when the appropriate sequence of IR pulses was detected. Explain how your system works. In what ways would this be better and worse than a mechanical lock?

Revision questions for Section F

1 A Morse code transmission system which uses radio waves contains the following functional blocks. Switch, buzzer, transmitting aerial, RF oscillator, filter, detector, receiving aerial, RF amplifier.
   a) Draw a suitable block diagram for the system.
   b) Explain how the whole system works.

2 State the approximate frequencies used for broadcasting
   a) AM radio on the medium wave band,
   b) FM radio on the VHF band,
   c) TV on the UHF band.

3 Explain how a communications satellite can be used to transmit signals from one radio station to another. What frequencies are used for satellite communications? Give one reason why.

4 Explain the meaning of the following terms used to describe how TV systems work:
   pixel, line scan, field scan, video signal,
   line sync pulse, field sync pulse.

5 Draw a circuit diagram to show how a transistor can be used to amplify AC signals. Don't bother about component values. Explain the function of the coupling capacitors and the bias network. In what way is an FET amplifier superior to a bipolar transistor one?

6 Figure F.1 shows a block diagram of an AM radio receiver.
   a) What does the term AM mean?
   b) Name the components labelled X, Y and Z.
   c) Describe the function of X.
   d) What is the function of the components Y and Z?
   e) What does the RF amplifier do ? Why should it use FETs?
   f) What does the detector do ? Illustrate your answer with sketches of typical waveforms going into it and coming out of it.
   g) Show how a detector can be made from a diode, a capacitor and a resistor.
   h) The system of figure F.1 has no provision for a volume control to adjust the level of sound fed out of the speaker. Draw a diagram to show how a potentiometer could be inserted to act as a volume control.

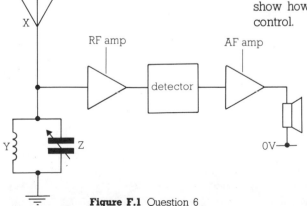

Figure F.1 Question 6

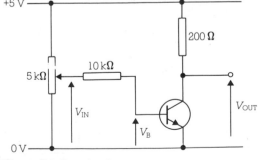

Figure F.2 Question 8

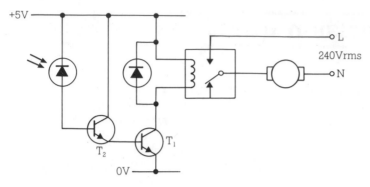

Figure F.3 Question 9

7 Radio telephones allow people to send and receive telephone calls from almost any location in the country. They are expensive to buy and expensive to use, compared with normal cord telephones. Why would it nevertheless be an advantage for the following people to use such a telephone system?
a)   A doctor working as a local GP.
b)   The general manager of a large company.

8 Figure F.2 shows part of an AC amplifier based on an npn transistor. The transistor has a current gain of 100. When V_{IN} is +5 V, the transistor is saturated.
a)   What is the value of V_{OUT}?
b)   How big is the collector current?
c)   What is the smallest base current that will keep the transistor saturated?
d)   What is the value of V_B when the transistor is saturated?
e)   What is the smallest value of V_{IN} that will keep the transistor saturated?
f)   What range of values of V_{IN} will turn the transistor off?
g)   What is the value of V_{OUT} when the transistor is off?
h)   Show how the addition of a couple of capacitors will convert the circuit of figure F.2 into an AC amplifier. Show where the signal to be amplified is fed in and where the amplified signal is fed out.
i)   Explain how you would use a voltmeter to help you adjust the bias to get a good operating point.
j)   What would happen to the amplified signal if the bias was incorrectly set?

9 The circuit of figure F.3 contains a photodiode which responds to infra-red light.
a)   Describe what happens to the photodiode when infra-red light
   i)   does not hit it,   ii)   hits it.
b)   The transistors have current gains of 50. The relay coil requires a current of 75 mA before it can pull the contacts over. For the motor to come on, how much current has to go through the
   i)   collector of T_1,   ii)   base of T_1,   iii)   collector of T_2
   iv)   base of T_2,   v)   the photodiode?
c)   Draw the circuit diagram of a system which could be used to provide infra-red light each time that a switch was pressed. Explain how it works.

30
The impact of electronics

Electronics is responsible for what is perhaps the most rapid period of change in human history. This chapter is going to summarise why electronics is steadily taking over other technologies and describe some of the consequences of this.

Advantages of electronics

There are several reasons why electronics is eventually going to be the dominant technology used by our society. For better or for worse, the electronic way of doing things is probably going to take over from any other way of doing the same thing.

Cheapness

Solid-state electronic devices can be very cheap to manufacture. ICs can be made in large numbers by a largely automatic process, so very little expensive human intervention is required for their fabrication.

Of course, the research and tooling costs involved in the manufacture of ICs are very large. It costs a lot of money to design them and factories for making them are expensive to build. However, if enough ICs are sold, only a small fraction of their purchase price is needed to cover these initial costs.

Large volume production of many ICs is ensured because of their versatility. Devices like CPUs and memories can be used to make a wide variety of systems. So the same IC can be used in many ways.

It is not only the components of electronics that are cheap. Because circuits may be constructed and tested by robots it is possible to produce whole circuit boards very cheaply. This degree of automation is only worthwhile if demand for a particular circuit board is high enough. So a popular microcomputer board might be made and tested completely automatically. A similar system, involving the same components but made by hand in only small numbers for a specialised role in the defence industry, might cost ten times as much.

Reliability

Electronic systems can be extremely reliable. Once a component has been tested after manufacture it is very unlikely to fail if it is used in a well-designed circuit. A well crafted piece of electronics needs no servicing, has no moving parts to wear out and can be very robust. Electronic systems age very slowly, so their characteristics change very little from year to year. It is possible, although not always profitable, to design electronic systems which protect themselves from overheating, being short-circuited and being destroyed by a faulty power supply.

Economy

Electronic components use up very little in the way of raw materials and need very little energy. The basic raw materials for ICs are sand and aluminium, both of which are plentifully available. The fact that calculators are routinely powered from the room lights via solar cells shows how little energy modern computers require! So the fuel costs of electronic technology can be minimal.

Compactness

Electronic systems can be miniaturised. A system which consists of several interconnected ICs on a circuit board can usually be replaced by a single IC. So electronics can be very compact.

Here are some examples. Most of the space in a digital watch is taken up by the battery. Indeed, a digital watch can have **any** shape, not just the classic disc shape. The electronics of a powerful microcomputer can be fitted into the case of its monitor, or underneath its keyboard. Most of a calculator is taken up by the battery, the display and the keyboard. A camera has its electronics stuffed into any spare corner of its case.

Speed

Electronic systems can operate at much higher speeds than alternative systems. A mechanical calculator at full speed cannot manage much more than one operation a second, whereas an electronic computer can accomplish over 100 000 per second with ease.

Sensors

A wide variety of sensors are available which can be used to transfer information in or out of an electronic system. Electronic systems can sense, among other things, visible light, pressure, temperature, infra-red light, radio waves, fluid flow, speed, position and angle. They can control almost anything via electric motors and they can emit a wide variety of waves, including radio, sound and light.

> **Electronics has several important advantages over other technologies. It can be cheaper, more reliable, less wasteful, smaller, faster and can respond to a wider range of signals.**

Inevitable consequences

Electronic devices are replacing mechanical ones even though they may not really work any better. For example, a digital watch is a great threat to clockwork ones because it can be produced more cheaply and not because it keeps better time. So mechanical controllers for washing machines and central heating systems will eventually be replaced by electronic ones.

Electronic systems are easy and cheap to repair. Electronics is a throwaway technology. If part of a system fails the whole lot is junked and replaced. It is far cheaper to throw away a computer circuit board and replace it within seconds with a new one than to try and locate the faulty component.

Electronic systems can be very large and complex. Enormous computers can be built without worrying that they will fail regularly because components stop working.

Electronic systems can process information rapidly. They can acquire, process, store and transmit enormous quantities of data quickly, cheaply and reliably. Computers, not libraries, are going to store information in the future. Information stored in them can be accessed much more rapidly than is possible with paper. So police can obtain information about a car that they are following within seconds of reaching for their radio microphone.

Electronics can extend our senses and capabilities. Electronic systems can be made to respond to touch, voice, smell and sight. They can be made more sensitive than our senses, and extend the range of things to be sensed. Electronic sniffers can smell out explosives or drugs hidden in airplanes. Satellites can explore the earth's surface for us, looking for minerals or studying the weather.

The pace of change

Electronics is developing new devices all the time. They are mostly bigger and better versions of previous devices. Thus several types of integrated circuit logic gate have come on the market over the years. They all look the same from the outside, but work in completely different ways inside. As a rule of thumb, electronics gets cheaper every year, allowing it to make progress into areas where the same function could be previously accomplished by other means.

Sometimes new electronics can create a whole new market, resulting in rapid expansion. The invention of the microprocessor allowed the cheap hand-held calculator to become a reality a few years later. It didn't replace anything, so it was immediately adopted. Similarly, the home computer was a new market. Once sales of home computers started to rise, the computers themselves came down in price, boosting sales even further. The powerful home computer was a familiar reality only five years after the first one was introduced.

Not all new electronic technology takes off quickly.

Digital watches were able to eventually dominate the market because mechanical watches have a limited lifetime, but people did not purchase a digital watch if their old mechanical one was still satisfactory. Mechanical cash registers and weighing machines are slowly disappearing as electronic ones take over. The huge investment in the old telephone technology has meant that modern electronic telephone exchanges are only slowly replacing the old electromechanical ones. Optical fibres will only be used widely when the old cables that they are going to replace have come to the end of their useful life.

Electronic scales are rapidly replacing mechanical ones because they can instantly tell the customer the cost of the goods being weighed out. *(Avery)*

Evils of electronics

The progress of electronics into many areas of life is inevitable, mainly for the economic reasons outlined above. Not all of the consequences of the widespread adoption of electronic technology will be seen as beneficial. Here are some of them.

Electronics is a secret technology. Because electronics has no moving parts and its functional components are minute, its operation is not obvious to the majority of its users. Previous technologies have been accessible to the general public because of their moving parts or wires. People could follow the chain of cause and effect with their eyes and hence feel that they understood what made the system tick. For this reason, electronic systems may be perceived as alien and unfriendly.

The use of electronics in factories to automate production of objects is making workers redundant. Companies which resist the introduction of a new technology which allows them to produce things better and more cheaply than before are going to eventually go bankrupt.

Electronics can be used to spy on people and invade their privacy. Word processors can measure the speed at which their operators are typing. Computers can store information about individuals without their knowledge. Miniature radio transmitters (bugs) can eavesdrop on peoples conversations. Hidden TV cameras can watch people and record their movements.

Electronics is extensively used for military purposes. Nuclear missiles rely on electronics to guide them to their target. Radar systems rely on electronics to identify possible enemy targets. Battlefield communications systems rely heavily on radio technology.

**Like all other technologies, electronics
can solve problems as well as create them.**

Blessings of electronics

The existence of electronic technology allows us to do many things which were hitherto impossible. Here are some examples.

Electronics can increase safety. Electronics can be used to give humans more information about systems (such as cars, planes, power stations or ships) than hitherto. Because electronics is cheap, you can afford to build in duplicate control systems so that if one fails, the other takes over automatically. Electronics allows planes to land safely in the fog, allows ships to know where they are at all times and can prevent cars from skidding.

Electronics can take over boring repetitive or dangerous tasks. There should be no need for humans to be the slave of a machine, making the same device day after day. Electronics can operate in hazardous environments such as coal mines, chemical factories and underwater.

Electronic computers allow us to process and store information much more efficiently than before. Actually, computers have been around for the last thirty years, but they have been relatively expensive for most of that time. Now that they are cheap, their use is becoming widespread. Our society is becoming increasingly complex and difficult to run as its population grows. It may well be that such a society is ungovernable without computers to deal with the flow of information (such as tax forms, car licences and social security claims) that government requires!

Electronics can do much to improve the quality of life of the disabled. Hearing aids for the deaf were one of the first applications of transistors.

A special electronic system allows this handicapped person to use a computer simply by moving his head. This allows him to print messages on the screen, write letters, do homework or to do any of the many things that computers can be used for. *(Rex Features)*

Electronics can educate. Computers can make very patient teachers, and TV is widely used for informing people about events and happenings throughout the world.

Electronics can entertain The popular music industry relies heavily on electronics for its products. Video recorders, synthesisers and cassette tape recorders are only possible through electronics.

Electronics can allow greater security. Electronic burglar alarms can alert the police automatically and tell them which room the burglar is in. At a national level, electronics allows a country to continuously survey the activities of its neighbours via radar or satellite.

Electronics allows people to communicate with each other over great and small distances. The world-wide telephone network is only possible because of the use of electronics. Radio communications are indispensible for the safety of people who live in isolated places. Radio contact with satellites allows man to explore the solar system. Air travel would be a hazardous affair if communication between planes and airports was not possible.

Electronics is a technology which promises to do a great deal for us. If it is used wisely it should solve many more problems than it creates.

Answers to questions

Using electricity p 7

1 a) B and C   b)   A and C

2 a)   b)   500 + 50 = 550 mA

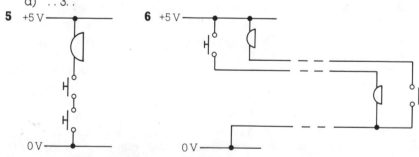

3 i)   80 mA   ii)   110 mA

4 a)   .. power supply rails .. power supply ..

    b)   .. output tranducers .. electrical ..

    c)   .. charge   .. conductor .. amps   .. ammeter .. insulators .. voltage .. voltage .. closed ..

    d)   .. 3 ..

5   **6**

Controlling current p 14

1 0.5 mA, 23 mA, 4.5 mA, 100 mA, 3.2 mA, 32 mA.

2 2.5 mW, 115 mW, 45 mW, 1000 mW, 48 mW, 480 mW.

    1, 2, 3 and 5 could have a power rating of 250 mW.

4 Nearest preferred value is 680 Ω.

5 iii)

6 a)   22 kΩ b)   470 Ω c)   100 kΩ d)   3.3 kΩ

7 a)   Blue, grey, orange, silver

    b)   Brown, red, brown, gold

    c)   Red, red, red, silver

    d)   Orange, orange, brown, gold

8 a) .. anode .. cathode .. 2 V ..
b) .. tolerance ..
c) .. series .. current through .. low ..
d) .. parallel .. voltage drop across .. large ..
e) .. three .. wiper .. track terminals ..

Controlling voltage p 19

1 X, Y and Z
2 8, 7, 5, 3 and 2
3

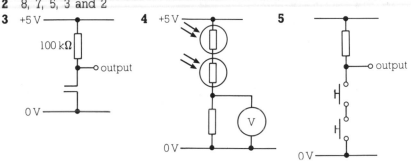

6 a) high .. low ..
b) .. voltage .. pull-up .. pull-down ..
c) .. light intensity .. low .. high ..

Increasing the current p 27

1 a) When A is low, Q is high; when A is high, Q is low.
b) ii)
2 a) i)

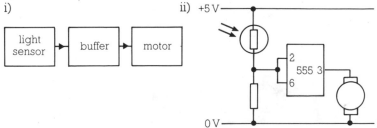

b) i)

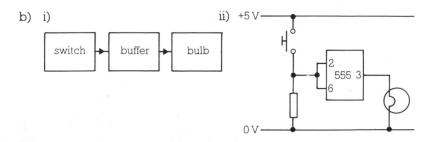

c) i)   ii)

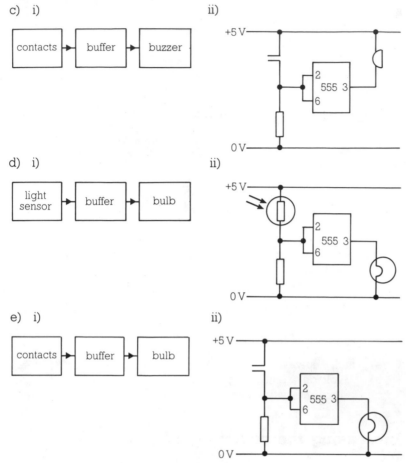

d) i)   ii)

e) i)   ii)

3 a) and c) by a 555 IC, the rest by a relay.
5 a)

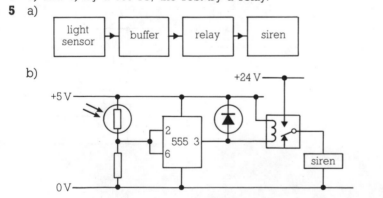

b)

Basic logic gates p 33

3 i)

B	A	C	Q
0	0	1	0
0	1	0	0
1	0	1	1
1	1	0	0

ii)

B	A	D	Q
0	0	1	0
0	1	1	1
1	0	0	0
1	1	0	0

iii)

B	A	E	F	Q
0	0	1	1	1
0	1	0	1	0
1	0	1	0	0
1	1	0	0	0

4

B	A	Q
0	0	0
0	1	1
1	0	1
1	1	1

B	A	C	Q
0	0	1	1
0	1	0	0
1	0	1	1
1	1	1	0

B	A	D	Q
0	0	1	1
0	1	1	1
1	0	0	0
1	1	1	0

B	A	E	F	Q
0	0	1	1	1
0	1	0	1	1
1	0	1	0	1
1	1	0	0	0

5

B	A	C	D	Q
0	0	1	1	0
0	1	0	1	0
1	0	1	0	0
1	1	0	0	1

AND

B	A	E	Q
0	0	1	0
0	1	1	0
1	0	1	0
1	1	0	1

AND

B	A	F	G	Q
0	0	1	1	0
0	1	0	1	1
1	0	1	0	1
1	1	0	0	1

OR

B	A	H	Q
0	0	1	0
0	1	0	1
1	0	0	1
1	1	0	1

OR

6

B	A	C	D	E	Q
0	0	1	1	1	0
0	1	0	1	0	1
1	0	1	0	0	1
1	1	0	0	0	1

OR

B	A	F	G	H	Q
0	0	1	1	1	0
0	1	0	1	1	0
1	0	1	0	1	0
1	1	0	0	0	1

AND

7

C	B	A	D	Q
0	0	0	0	0
0	0	1	0	0
0	1	0	0	0
0	1	1	1	0
1	0	0	0	0
1	0	1	0	0
1	1	0	0	0
1	1	1	1	1

Q is only 1 when all three inputs are 1.

C	B	A	E	Q
0	0	0	0	0
0	0	1	1	1
0	1	0	1	1
0	1	1	1	1
1	0	0	0	1
1	0	1	1	1
1	1	0	1	1
1	1	1	1	1

Q is only 0 when all three inputs are 0.

Designing logic systems p 39

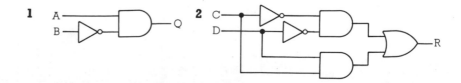

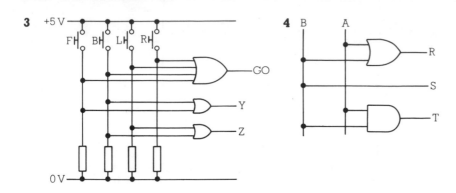

3

4

Doing it with NAND gates p 42

1   .. 1 .. 0 .. NOT ..

2

A	F	Q		B	A	C	Q		B	A	D	E	Q
0	1	0		0	0	1	0		0	0	1	1	0
1	0	1		0	1	1	0		0	1	0	1	1
				1	0	1	0		1	0	1	0	1
				1	1	0	1		1	1	0	0	1
	wire				AND					OR		

3 a)

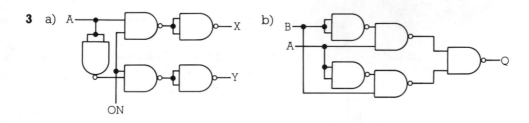

b)

c)

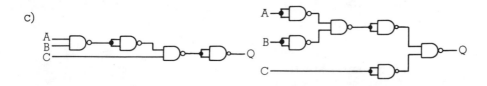

d)

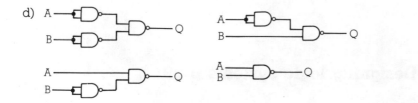

Revision questions for section A p 43

1 a) Parallel   b) i) R   ii) L

c)

switch L	switch R	ammeter
open	open	0 mA
open	closed	20 mA
closed	open	80 mA
closed	closed	100 mA

d) Bulb. Its power is 6 x 80 = 480 mW.

2 a)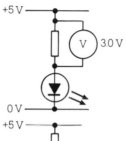

b) 15 mA
c) 220 Ω; red, red, brown, gold. Power rating of 250 mW.
d) 6.4 mA, dimmer.
e) 19 mW, no danger.
g) Become zero.

f) (circuit with ammeter)

3 a) When A is above +2.5 V, Q is 0 V. When A is below +2.5 V, Q is +5 V.

b) Light dependent resistor. Resistance decreases as light intensity increases.

c) Shine lots of light on it. About 5 V.

d) i) B lit, T out.   ii) B out, T lit.

e) In the dark, B is lit. In the light T is lit.

4 a)

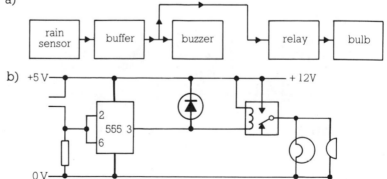

b) (circuit with 555)

5 a) Gate 1 is an OR gate; its output only goes low when both its inputs are low.
Gate 2 is an AND gate; its output only goes high when both of its inputs are high.
Gate 3 is a NOT gate; its output is high when its input is low and low when its input is high.

Gate 1	B	A	Q
	0	0	0
	0	1	1
	1	0	1
	1	1	1

Gate 2	B	A	Q
	0	0	0
	0	1	0
	1	0	0
	1	1	1

Gate 3	A	Q
	0	1
	1	0

b) A 0 is any voltage between 0 V and +2.5 V. A 1 is any voltage between +2.5 V and +5.0 V.

c)

L	R	X	Y	B	A
0	0	0	1	0	0
0	1	1	1	0	1
1	0	1	1	0	1
1	1	1	0	1	0

d) i)   ii)   iii)

e)

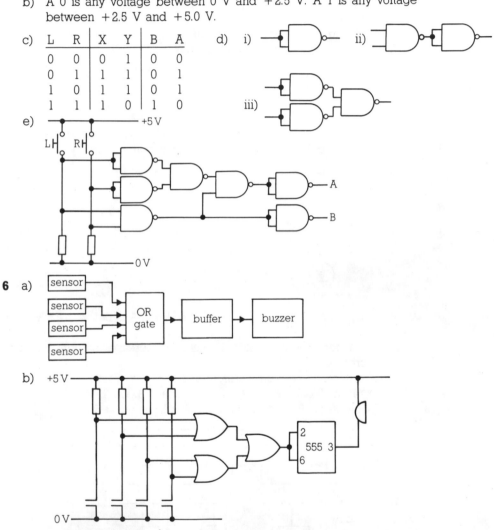

6 a)

b)

Capacitors p 51

1 a) i) X ii) Y
  b) The bulb glows brightly and then gradually fades out.
  c) The initial brightness of the bulb is as before, but it takes twice as long to fade completely out.

2 a) i) The motor rotates all of the time.
      ii) The motor rotates for a certain length of time and then stops.
  b) i) 10 s ii) 2.5 s iii) 20 s
  c) 300 kΩ, 100 μF

d)   Voltage/V

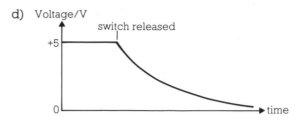

3 a) i) low   ii) high
   b) i) high   ii) low
   c)   ..1.. is off.. goes off.. comes on..5s..0.. comes on..

4   ..1..1..1.. time constant..*R*..*C*..0.. doubling.. halving

Relaxation oscillators p 61

1

	i)	ii)	iii)	iv)
a)	+2 V	+4 V	+10 V	+1.5 V
b)	−2 V	0 V	−10 V	0 V
c)	30 ms	3 ms	40 ms	6 ms
d)	33 Hz	333 Hz	25 Hz	167 Hz

2 a) 500Ω   b) 2.5 kΩ   c) 10 kΩ

3 a)

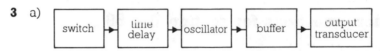

   b) 2.3 Hz
   c) i) Alternates between 5 V and 0 V, spending about ¼ s in each state.
      ii) Stays at 0 V.
   d) 10s   e) U   f) 20

4 a)

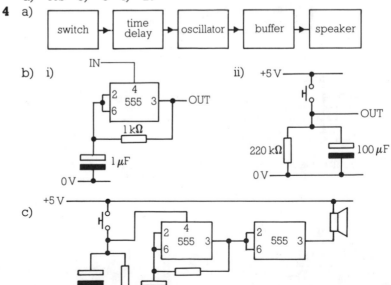

   b) i)   ii)

   c)

5 a) 2 V b) 30 ms c) timebase 10 ms/cm, vertical amplifier 1 V/cm.

Pulses p 67

1 a)

```
┌──────────┐   ┌──────────┐   ┌────────┐   ┌────────┐
│  light   │──▶│monostable│──▶│ buffer │──▶│ buzzer │
│  sensor  │   └──────────┘   └────────┘   └────────┘
└──────────┘
```

c)

3 a) i) High ii) Low
   b) i) Stays at 0 V ii) Stays at +5 V
       iii) Goes to +5 V for a while before returning to 0 V.
   c) 10 s
   d) Comes on and stays on until the switch is released.

Revision questions for section B p 69

1 a) Polarised or electrolytic b) +5 V
   c) Voltage/V
   d) i) 14 s ii) 3.5 s

```
Voltage/V
                released
  +5 ┤        ╮│
     │        │ ╲
     │        │  ╲___
     │  pressed      ‾‾‾‾───
   0 └─────┘─────────────────▶ time
```

2 a) 20 ms b) 30 ms c) 33 Hz
   d)

3 a) The bulb goes on and off, spending ½ s on and ½ s off.
   b) i) 100 Hz ii) Be on all the time at half brightness.
   c) see figure 9.14

4 a) Speaker
   b) A = 1 s, B = 5 ms   c)
   d) i) C = B ii) C = 0
   e) Each burst of sound lasts for
       ½ second and has a
       frequency of 200 Hz.

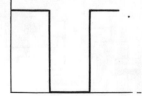

5 a) Lower T from +5 V to 0 V.
   b) Q goes to +5 V for 10 s before returning to 0 V.
   c) Double the capacitor to 200 μF or double the resistor to 200 kΩ.

6 a) Schmitt trigger NOT gate
   b) i) 4 V ii) 1 V c) i) 0.5 V ii) 4.5 V
   d) see figure 9.5

Bistables p 75

1 ..1..1..0..0..0..1..1

2 a) 0   b) 0   c) It doesn't make a noise
   d) It makes a noise   e) It still makes a noise
   f)

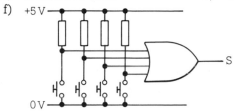

3 a)

B	A	Q
0	0	1
0	1	1
1	0	1
1	1	0

b)

$\bar{S}$	$\bar{R}$	Q	$\bar{Q}$
0	1	1	0
1	0	0	1

   c) i) Yes   ii) Yes

   d) i) Hold $\bar{R}$ at 1 and pulse $\bar{S}$ to 0.
      ii) Hold $\bar{S}$ at 1 and pulse $\bar{R}$ to 0.
   e)

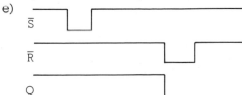

Remembering words p 85

1 a) see figure 12.6
   b) ..S..R..1..0..0..1..0..0..1..D..Q..rising..
   c)

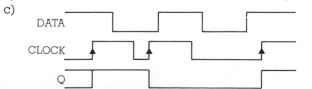

2   see figure 12.12

3   8   4   2   1   3
       9   7   0   6   5

4

0	1	2	3	4	5	6	7	8	9
0000	0001	0010	0011	0100	0101	0110	0111	1000	1001

5   see figure 12.15

6   see figure 12.16

7   Place the word to be stored at the three inputs C, B and A. Pulse CK high. The word will be frozen at X, Y and Z.

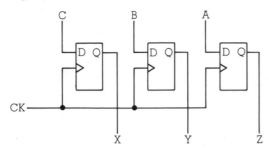

8   Convert each of the two digits of the number into a four bit binary word using BCD. Place each word at the inputs of the two ICs and pulse $\overline{ST}$ low.

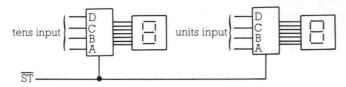

Semiconductor memories p 95

1   Address lines to select one of the memory stores in the system. Data lines for transmitting binary words in or out of the system. Control lines for instructing the system to store a word present on the data lines in a memory store or to put a word from a memory store onto the data lines.

2   see p 88

3   a)   Make A_1A_0 = 01. Hold $R/\overline{W}$ at 0. Make $D_3D_2D_1D_0$ = 1110. Pulse $\overline{CE}$ low.

   b)   Make A_1A_0 = 10. Hold $R/\overline{W}$ at 1. Hold $\overline{CE}$ low and look at $D_3D_2D_1D_0$.

4 a)

A	Q	O_1	O_0
0	0	0	0
0	1	0	1
1	0	0	0
1	1	1	0

b)

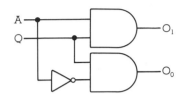

c)

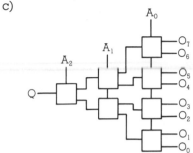

5 a) i) Press all three switches   ii) Green   iii) Any two

b)

$A_2A_1A_0$	$O_2O_1O_0$
000	000
001	001
010	001
011	011
100	001
101	011
110	011
111	111

9 a) Tristates. They disconnect the outputs of the memory from any system they are connected to when $\overline{CE}$ is 1.

b) Demultiplexer. Does the address decoding.

c)

A_1	A_0	Q_3	Q_2	Q_1	Q_0
0	0	0	0	0	1
0	1	0	0	1	0
1	0	0	1	0	0
1	1	1	0	0	0

d)

A_1	A_0	O_1	O_0
0	0	0	0
0	1	1	1
1	0	1	1
1	1	1	0

Revision questions for section C p 97

1 a) It has two stable output states for one input state. So H can be 1 or 0 if E = 0 and F = 0.

b) i) Hold E high and F low   ii) Hold F high and E low

c) It stays in whatever state it was in before.

d) i) Low ii) High

e) 1 and 2. The buzzer continues to work.

f) 3 or 4. The buzzer stays off.

2 a) D flip-flop   b) see figure 12.6   c) 0

d)

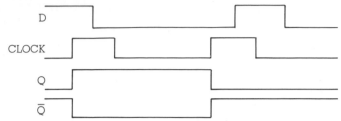

e) It can store one bit (a 1 or a 0). That bit can be written into the component as well as read out of it. So it behaves like a read-and-write memory.

3 a) Seven segment LED   b) i) 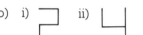   ii) ⎩⎧   iii) ⎫

c) BCD to seven segment decoder

d) i) Zero
    ii) Three
    iii) Nine

e) i) 0001   ii) 0111   iii) 0010

f) Each time that CK is raised from 0 to 1 the word $D_3D_2D_1D_0$ is read in and frozen at $Q_3Q_2Q_1Q_0$.

g) see figure 12.14

5 a) 8   b) 2 bit

c) Make $A_2A_1A_0$ = 110. Hold R/$\overline{\text{W}}$ at 0. Hold D_1D_0 at 10. Pulse $\overline{\text{CE}}$ low for an instant.

d) Make $A_2A_1A_0$ = 010. Hold R/$\overline{\text{W}}$ at 1. Hold $\overline{\text{CE}}$ low and look at D_1D_0.

e) Random-access-memory, but really means read-and-write memory.   i) no change   ii) lost forever.

Operational amplifers p 105

1 a) .. −4V . +4 V ..

b) .. the inverting one .. the non-inverting one ..

c) .. split .. +5 V .. −5 V ..

2 a) +3.4 V

b) i) +4 V   ii) −4 V   iii) −4V   iv) −4V

c) Between +5 V and 3.4 V   d)

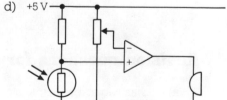

3 0.22 V   0.45 V   0.88 V   4.12 V   4.54 V

4 a) i) 2.4 kΩ   ii) 3.9 kΩ   iii) 5.6 kΩ

b) + 1.0 V   c) 1.0 V

d) The voltage drops from 1.5 V to 0.75 V. Initially the buzzer is off.

As soon as the voltage goes below 1.0 V the buzzer comes on and stays on.

5 −4 V   +4 V

7 Between +2 V and +3 V.

Audio systems p 113

1 a) .. alternating .. speaker ..

      b) .. 10 mV .. 100 kΩ .. 8 Ω .. 100 mW ..

      c) 16 Hz .. 16 kHz ..

      d) .. amplifier .. amplitude of output ÷ amplitude of input ..

      e) .. coaxial cable .. mains electricity ..

2 a) Voltage amplifier because gain is greater than 1 and output current very low.

      b) see figure 15.7   c) 1 V

      d) Yes because of its high input resistance which will not draw much current from the microphone.

      e) Doesn't amplify over the whole audio frequency range.

      f) Overheat and self-destruct in flames!

      g) If run off 15 V supply rails, maximum amplitude of output unlikely to be greater than $1/2 \times 15 = 7.5$ V. If gain is 40, then maximum amplitude of distortion-free input signal is $7.5 \div 40 \simeq 0.19$ V.

Voltage amplifiers p 120

1 b)

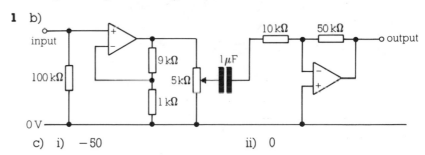

      c) i) −50   ii) 0

3

	a)	b)	c)
i)	inverting	−10	−2.5 V
ii)	non-inverting	+5	−0.5 V
iii)	inverting	−0.1	−0.5 V
iv)	non-inverting	+4	+0.8 V
v)	inverting	−50	+0.5 V
vi)	non-inverting	+22	−2.2 V

4 a)   b)

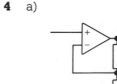
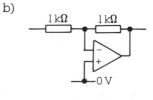

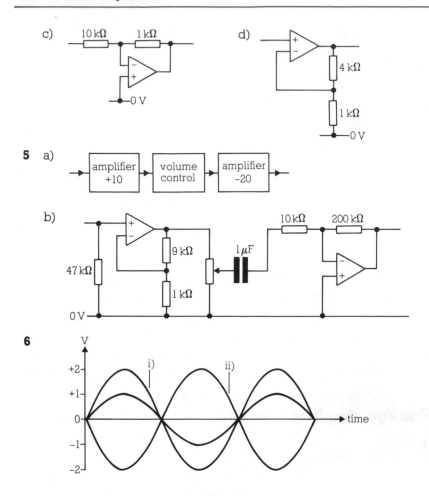

6

Power amplifiers p 129

1 a) see figures 17.4, 17.6
   b) .. 0.7 .. emitter .. collector .. base .. 0.7 .. emitter .. collector ..
       base.

2

	a)	b)	c)	d)
i)	2.0 V	9 mA	9 mA	0.09 mA
ii)	7.7 V	8 mA	8 mA	0.08 mA
iii)	4.0 V	40 mA	40 mA	0.40 mA

3 a) see figures 17.8, 17.9
   b) .. 0.7 .. collector .. 0.7 .. emitter .. base ..

4

	a)	b)	c)	d)
i)	− 2.0 V	4.2 mA	4.2 mA	0.021 mA
ii)	− 6.7 V	30 mA	30 mA	0.15 mA
iii)	− 11.3 V	113 mA	113 mA	0.57 mA

5 see figures 17.13, 17.14. Maximum amplitude = 7.3 V.

6

V_{IN}/V	S	T	V_{OUT}/V
+8.7	on	off	+8.0
+2.0	on	off	+1.3
+0.5	off	off	0.0
−0.3	off	off	0.0
−3.3	off	on	−2.6
−5.7	off	on	−5.0

7 a)

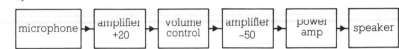

b) see figures 17.13, 16.9 and 16.2.

Simple power supplies p 141

2 .. live .. alternating .. live .. neutral .. current .. switch .. live .. neon .. live .. neutral .. the metal casing ..

3 0.5 A. See figure 18.8

4 a) 0.7 V   b) 5.7 V   c) 4.0 V   d) It would halve

5

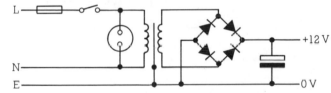

6 a) 8.5 V   b) 7.8 V   c) −7.8 V

d)

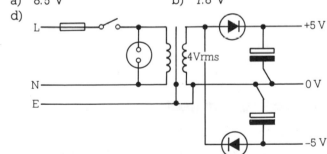

7 .. reverse .. very large .. no .. forward .. low .. blow .. isolate

8

	a)	b)	c)	d)	e)	
T = +10 V	V	+9.3 V	X	+0.7 V	Y	→ Z
B = +10 V	U	+9.3 V	W	+0.7 V	Y	→ Z

Regulated power supplies p 146

1 a) 143 mA   b) 22 Ω

  c) V_{OUT}/V

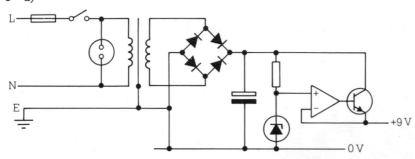

2 see p 143
3 a)

  b) 680 Ω
  c) The op-amp compares the output voltage with that across the
      zener diode (a fixed 9.1 V, provided that the mains voltage
      doesn't drop too far). If it finds any difference between them it
      adjusts the voltage at the transistor's base until that difference
      disappears.
  d) 10 000 μF because it will mean less ripple on the op-amp
      power supply rails.

Revision questions for section D p 147

1 a) A is a microphone. It generates alternating voltages when sound
      waves hit it. B is an amplifier. If feeds out an enlarged copy of the
      waveform which is fed into it. C is a speaker.
  b) The signal fed out of it has an amplitude that is 20 times larger
      than the amplitude of the signal which was fed into it. Both
      signals have the same sign of voltage as each other.
  c) A 1 kHz sine wave with an amplitude of 400 mV.
  d) An intercom?
2 a) ..1 V.. −11 V.. +11 V..   b)   +8 V
  c) i) < +8V   ii) > +8V   d)   2 kΩ
  e) $R = (11 - 2) \div 5 = 1.8$ kΩ
3 a) i) Goes to +5 V immediately
      ii) Slowly drops back to 0 V
  b) i) > +2 V   ii) < +2 V
  c) i) increases   ii) decreases

4 a) non-inverting   b) +5

V_{IN}/V	0	+1	+2	+3	−1	−2
V_{OUT}/V	0	+5	+10	+11	−5	−10

5
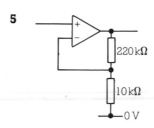
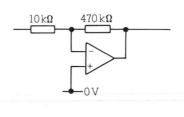

6 a) 16 Hz to 16 kHz
   b) The first section is an inverting amplifier which increases the amplitude of the signal fed into it. The second section is a power amplifier which provides enough current for the amplified signal to drive a speaker.
   c) −50   d) i) 1 V   ii) 1 V   e) T_1. T_2.
   f) see figure 15.8   g) i) No   ii) Yes
7 See figure 17.4 . . . 0.7 . . emitter . . less than 0.7 V above the emitter . . collector . . emitter . .
8 a) i) 50 mA   ii) 0.5 mA   iii) ≈50 mA
   b) i) 2.0 V   ii) 2.7 V   iii) +5 V
   c) see figure 17.6
11 i)   ii)   iii)

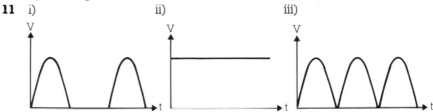

12 a) see figure 18.8
   b) live = brown, neutral = blue, earth = yellow and green.
   c) S is a zener diode. When it is reverse biased the voltage across it remains roughly constant regardless of the current going through it.
   d)

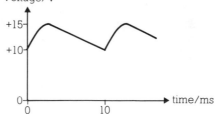

   e) i) no effect   ii) smaller than before
13 See the answer to Q 3 on p 146
14 a) 4 V   b) 40 mA   c) 25 mA   d) 40 mA
   e) see figure 19.7

Binary counters p 158

1 a) i) Changes state   ii) Doesn't change state
   b) If IN falls from 1 to 0, CK rises from 0 to 1 because of the NOT
       gate in between. Since $\overline{Q}$ always has the opposite state to that of
       Q, that rising edge makes Q the opposite of what it was before.
   c)
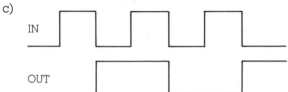

2 31, 3
3 a) see figure 20.7   b) BA = 00
   c)
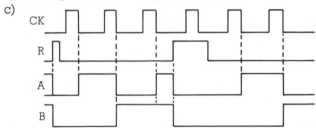

   d)
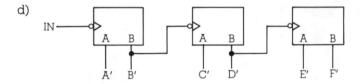

4 a) 5   b) i) 1000   ii) 1111
5 a)

   b)

Waveform	CK	A	B	C	D
Period/ms	¼	½	1	2	4
Frequency/kHz	4	2	1	½	¼

Decimal counters p 164

1 a) DCBA = 0101 → 0000

b)

Pulse number	D	C	B	A	R
0	0	0	0	0	0
1	0	0	0	1	0
2	0	0	1	0	0
3	0	0	1	1	0
4	0	1	0	0	0
5	0	1	0	1	1
	0	0	0	0	0
6	0	0	0	1	0

c)

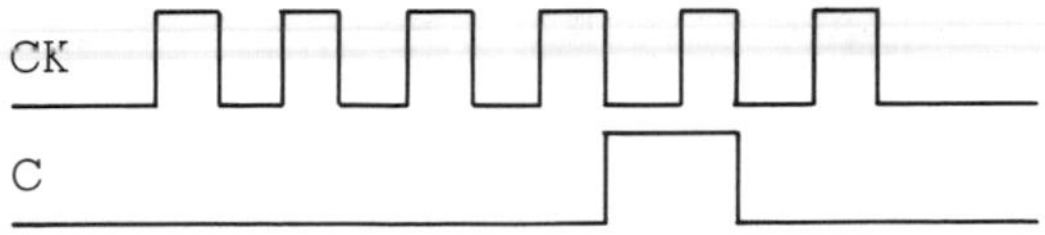

2 i)

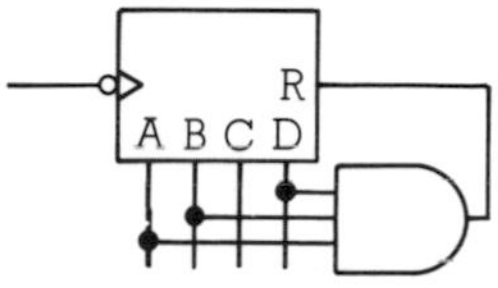

ii)

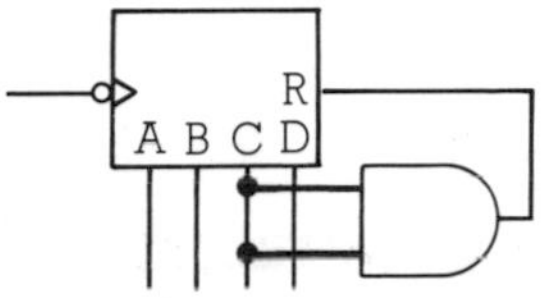

iii)

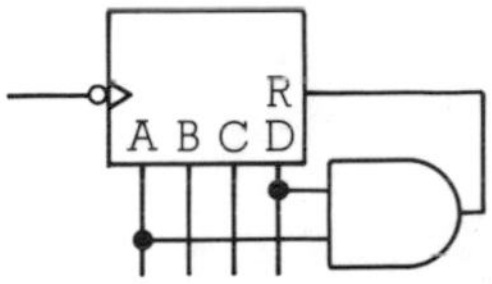

iv)

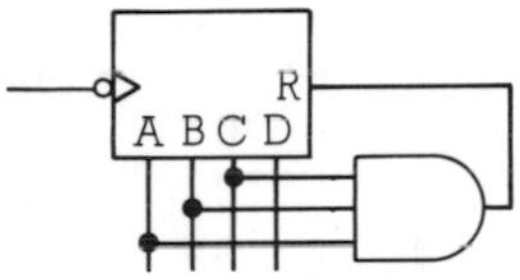

v)

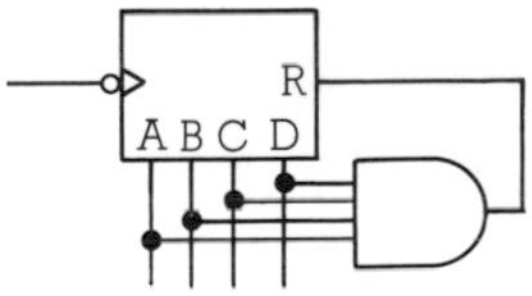

3

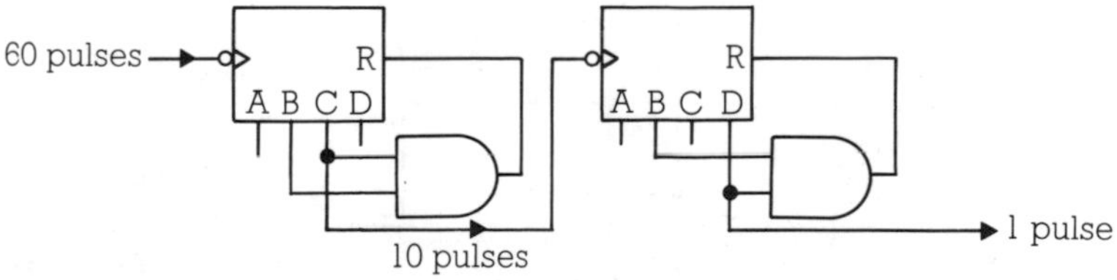

4

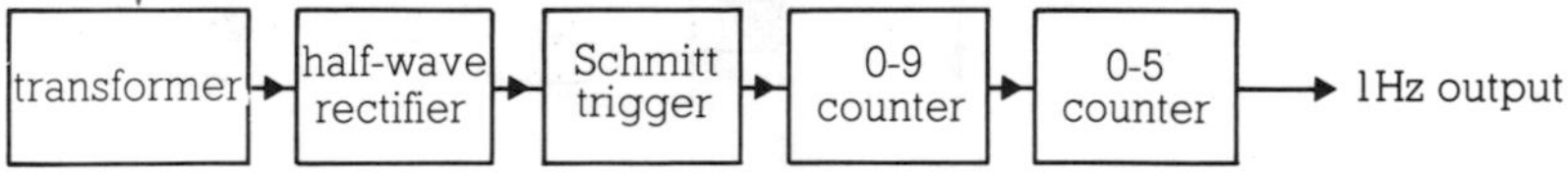

5 a) 111   b) 001

c)

Pulse number	C	B	A
0	1	0	0
1	1	0	1
2	1	1	0
3	0	0	1
4	0	1	0
5	0	1	1
6	1	0	0

6 a) DCBA = 0111 or 1111   b) 0011
    c) i) The numbers from 0 to 6 being displayed in rapid succession.
        ii) A number between 0 and 6 frozen on the display.
    d)

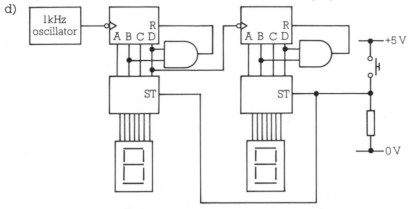

Sequence generators p 172

1 a)

B	A	X	Y	Z
0	0	1	0	0
0	1	0	1	0
1	0	0	0	1
1	1	0	1	0

b) The LEDs come on one at a time in the following sequence; top, middle, bottom, middle.

2

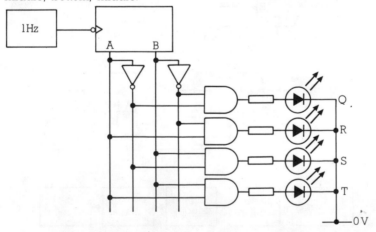

Microprocessor systems p 179

1 see figure 23.4

Digital-to-analogue converters p 185

2 a) -0.61 V   b) 0.0 V   c) $+0.30$ V

3 a) OA_1 is a summing amplifier, OA_2 is an inverting amplifier.
   b) 1.6 kΩ   c) 5 kΩ, 20 kΩ and 40 kΩ
   d)

DCBA	V_{OUT}/V
0000	0.0
0001	0.2
0010	0.4
0100	0.8
1000	1.6

4

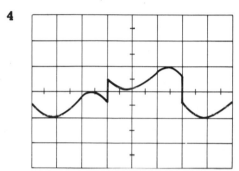

5

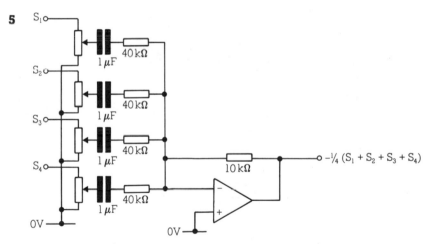

6 a)   see figure 24.2

Analogue-to-digital converters p 191

2 a)

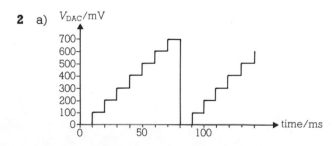

b) It displays the numbers from 0 to 7 in order, changing 100 times a
     second, so all you see is the number 8.
c) see figure 25.3
e) i) 3   ii) 6   iii) 7 probably
3 a) i) +4 V   ii) +4.5 V
b) i) −4 V   ii) +0.5 V

c)

Input signals	Output signal/V
$V_+ > V_-$	+3.3
$V_+ < V_-$	0.0

Revision questions for section E p 193

1 a) i) 1   ii) 0
b) Make P rise from 0 to 1.
c) see figure 20.6
2 a) 000   b) CBA stays at 000 and doesn't change.   c) Five

d)

Pulse number	C	B	A
0	1	0	0
1	1	0	1
2	1	1	0
3	1	1	1
4	0	0	0

e) see figure 20.10
3 a) DCBA = 0000 → 0001 → 0010 → 0011 → 0100 → 0101 → 0110 → 0111
     → 1000 → 1001 → 0000.

b)

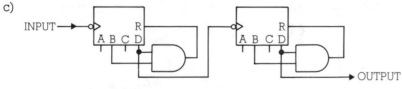

c)

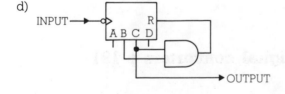

d)

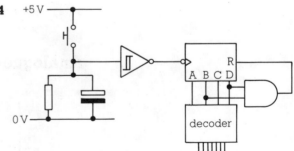

5 a)

B	A	E	F
0	0	0	0
0	1	0	1
1	0	0	1
1	1	1	1

b)

Time/s	0-10	10-20	20-30	30-40	40-50	50-60
Red	off	off	off	on	off	off
Green	off	on	on	on	off	on

10

CBA	V_{SUM}/V	V_{DAC}/V
000	0.0	0.0
001	−0.125	+0.5
010	−0.25	+1.0
100	−1.0	+4.0

11 b) i) 011   ii) 111   iii) 001

c)

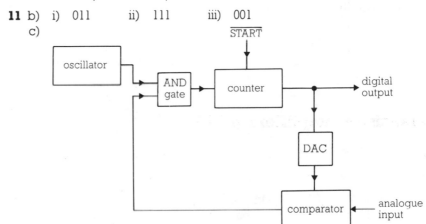

Radio p 206

1 a) 300 MHz to 3 GHz   b) 3 MHz to 30 MHz
c) 300 kHz to 3 MHz   d) 30 MHz to 300 MHz

2

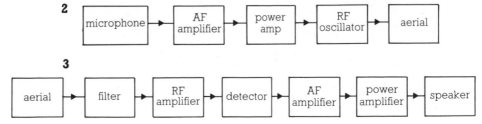

3

aerial → filter → RF amplifier → detector → AF amplifier → power amplifier → speaker

4 With amplitude modulation the amplitude of the radio carrier is changed by the signal which is to be carried by it. With frequency modulation, it is the frequency of the carrier which is changed by the signal.

5 a) Inductor, variable capacitor, germanium point-contact diode, aerial, capacitor.
   b) B
   c) .. aerial .. alternating .. resonant frequency .. tuned .. alternating .: rectified .. smoothed.
   d)

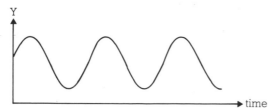

   e)

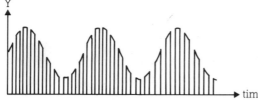

   f) Remove the headphones and replace with an audio amplifier, power amplifier and speaker.

Transistor amplifiers p 214

1 a) 0 V b) 2 mA c) 0.02 mA d) 2.0 V e) 2.7 V

2 a) not glow   b) i) 50 mA ii) 0.5 mA

   c) $R = 4.3 \div 0.5 = 8.6$ kΩ for the transistor to be saturated so 10 kΩ will not let enough current into base, so use 4.7 kΩ.

3 a)

V_{IN}/V	0.0	1.7	2.7	3.7	4.7
I_B/mA	0.0	0.1	0.2	0.3	0.4
I_C/mA	0.0	5.0	10.0	12.5	12.5
V_{OUT}/V	5.0	3.0	1.0	0.0	0.0

   b)

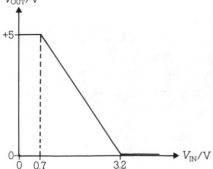

4 see figure 27.4

5 i) D ii) C iii) A iv) B

6 .. off .. zero .. zero .. 5 V .. on .. fall .. saturated .. base
.. collector ..

7 see figure 27.8

Television p 220

1 a) .. pixels .. video .. amplitude ..
   b) .. line sync .. scan .. start a new field scan ..
   c) .. 50 .. raster .. 0.04 ..

2 Include entertainment, education, news, advertising, sport, music, weather information and party political broadcasts.

Infra-red p 225

1 a) Infra-red LED. Emits IR light when the switch is pressed.
   b) Photodiode. The current through it increases as the amount of IR light hitting it increases.
   c) i) 4 V   ii) 0.018 mA   iii) 0.09 μA

2 a)

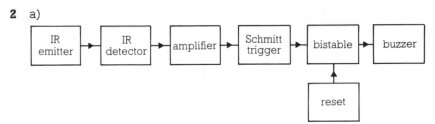

   b)

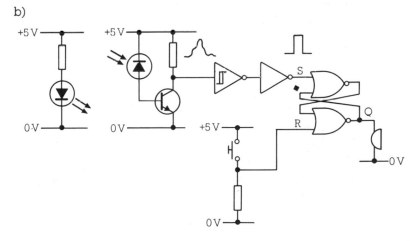

5

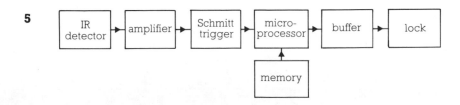

Revision questions for section F p 227

1 a) see figure 26.1

2 a) 300 kHz to 3 MHz
   b) 30 MHz to 300 MHz
   c) 300 MHz to 3 GHz

5 see figure 27.4

6 a) Amplitude modulation. The amplitude of the carrier wave is controlled by the signal being transmitted.
   b) Aerial, inductor, variable capacitor.
   c) It picks up the radio waves, converting them into alternating currents.
   d) Y and Z act as a filter. Only that alternating current in the aerial whose frequency is the same as the resonant frequency of the filter will generate an alternating voltage across the filter.
   e) The RF amplifier makes a larger copy of the alternating voltage across the tuned circuit. It should use FETs to cut down on the current drawn from the tuned circuit.

f)

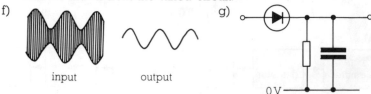

input   output

g)

h)

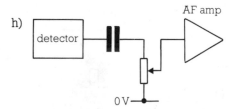

8 a) 0 V b) 25 mA c) 0.25 mA d) 0.7 V
   e) 3.2 V f) 0 V to 0.7 V g) 5 V h) figure 26.1
   i) Use a voltmeter to measure the collector voltage for no AC signal going into the amplifier. Adjust V_{IN} until the collector is at +2.5 V.
   j) distorted

9 a) i) no current   ii) current goes through
   b) i) 75 mA   c)
      ii) 1.5 mA
      iii) 1.5 mA
      iv) 30 μA
      v) 30 μA

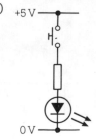

Appendix

Electronic circuit symbols

Resistor

Potentiometer

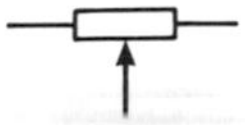

LED

Ammeter

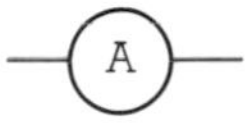

Push switch

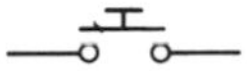

Relay

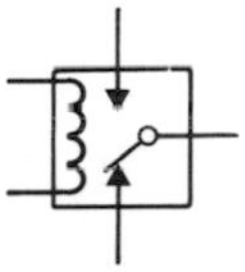

Bulb

npn transistor

Microphone

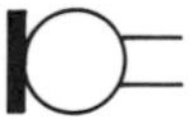

AND gate

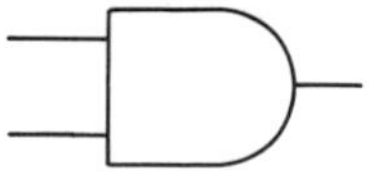

NAND gate

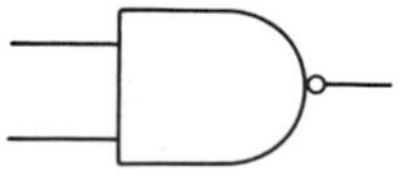

NOT gate

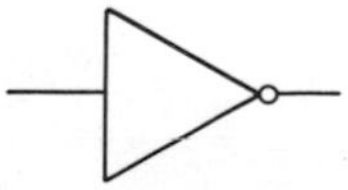

D flip-flop

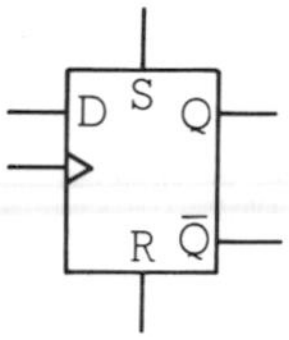

Aerial

Fuse

Zener diode

Capacitor

Diode

LDR

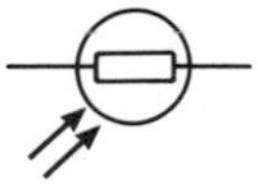

Voltmeter

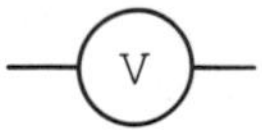

Switch

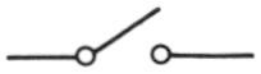

Thermistor

Inductor

pnp transistor

Loudspeaker

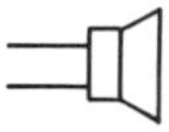

OR gate

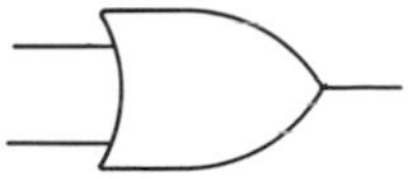

NOR gate

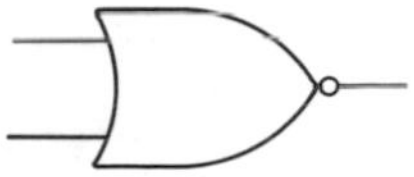

Op-amp

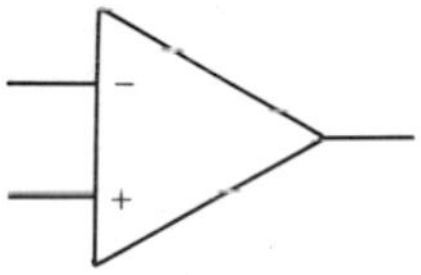

Binary counter

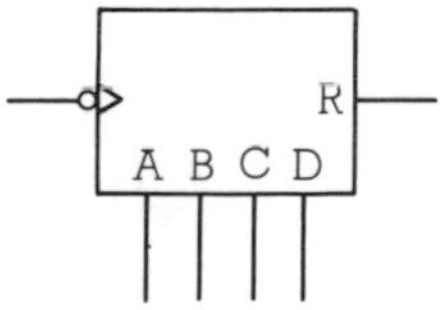

Transformer

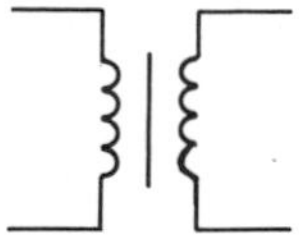

Neon bulb

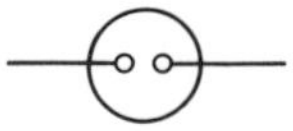

Contacts

Schmitt trigger
NOT gate

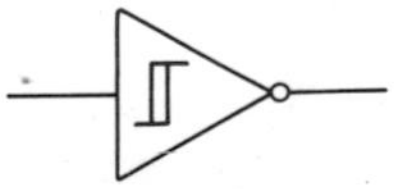